RF and Microwave
Wireless Systems

RF and Microwave Wireless Systems

KAI CHANG
Texas A&M University

A WILEY-INTERSCIENCE PUBLICATION

JOHN WILEY & SONS, INC.

NEW YORK / CHICHESTER / WEINHEIM / BRISBANE / SINGAPORE / TORONTO

This book is printed on acid-free paper. ∞

For ordering and customer service, call 1-800-CALL-Wiley.

Library of Congress Cataloging-in-Publication Data:

Chang, Kai, 1948–
    RF and microwave wireless systems / Kai Chang.
      p. cm.
    ISBN 0-471-35199-7 (cloth : alk. paper)
    1. Wireless communication systems. 2. Mobile communication systems. 3. Microwave systems. 4. Radio frequency. I. Title.

    TK5103.2 C45 2000
    621.3845--dc21

Printed in the United States of America.

10 9 8 7 6 5 4 3 2 1

To my parents and my family

Contents

Preface

Wireless personal mobile and cellular communications are expected to be one of the hottest growth areas of the 2000s and beyond. They have enjoyed the fastest growth rate in the telecommunications industry—adding customers at a rate of 20–30% a year. Presently, at least six satellite systems are being developed so that wireless personal voice and data communications can be transmitted from any part of the earth to another using a simple, hand-held device. These future systems will provide data and voice communications to anywhere in the world, using a combination of wireless telephones, wireless modems, terrestrial cellular telephones and satellites. The use of wireless remote sensing, remote identification, direct broadcast, global navigation, and compact sensors has also gained popularity in the past decade. Wireless communications and sensors have become a part of a consumer's daily life. All of these wireless systems consist of a radio frequency (RF) or microwave front end.

Although many new wireless courses have been offered at universities and in industry, there is yet to be a textbook written on RF and microwave wireless systems. The purpose of this book is to introduce students and beginners to the general hardware components, system parameters, and architectures of RF and microwave wireless systems. Practical examples of components and system configurations are emphasized. Both communication and radar/sensor systems are covered. Many other systems, such as, the global positioning system (GPS), RF identification (RFID), direct broadcast system (DBS), surveillance, smart highways, and smart automobiles are introduced. It is hoped that this book will bridge the gap between RF/microwave engineers and communication system engineers.

The materials covered in this book have been taught successfully at Texas A&M University to a senior class for the past few years. Half of the students are from RF and microwave areas, and half are from communications, signal processing, solid-state, optics, or other areas. The book is intended to be taught for one semester to an undergraduate senior class or first-year graduate class with some sections assigned to students for self-study. The end-of-chapter problems will strengthen the reader's knowledge of the subject. The reference sections list the principal references for further reading.

Although this book was written as a textbook, it can also be used as a reference book for practical engineers and technicians. Throughout the book, the emphasis is on the basic operating principles. Many practical examples and design information have been included.

I would like to thank all of my former students who used my notes in class for their useful comments and suggestions. I would also like to thank Mingyi Li, Paola Zepeda, Chris Rodenbeck, Matt Coutant and James McSpadden for critical review of the manuscript. Michelle Rubin has done an excellent job in editing and preparing the manuscript. Taehan Bae has helped to prepared some of the art work. Finally, I wish to express my deep appreciation to my wife, Suh-jan, and my children, Peter and Nancy, for their patience and support.

KAI CHANG

February 2000

Acronyms

AF	array factor
AGC	automatic gain control
AM	amplitude modulated
AMPS	advanced mobile phone service
APTS	advanced public transit systems
ASK	amplitude shift keying
ATIS	advanced traveler information system
ATMS	advanced traffic management system
AUT	antenna under test
AVCS	advanced vehicle control system
AVI	automatic vehicle identification
BER	bit error rate
BPF	bandpass filter
BPSK	biphase shift keying
BSF	bandstop filter
BW	bandwidth
CAD	computer-aided design
CDMA	code division multiple access
CEP	circular probable error
CMOS	complementary MOS
CP	circularly polarized
CPL	cross-polarization level
CRTs	cathode ray tubes
CT1/2	cordless telephone 1/2
CTO	cordless telephone O
CVO	commercial vehicle operations
CVR	crystal video receiver
CW	continuous wave

DBS	direct broadcast satellite
DC	direct current
DECT	digital european cordless telephone
DR	dynamic range
DRO	dielectric resonator oscillator
DSB	double side band
DSSS	direct-sequence spread spectrum
ECCMs	electronic courter- countermeasures
ECMs	electronic countermeasures
EIRP	effective isotropic radiated power
EM	electromagnetic
ESM	electronic support measure
EW	electronic warfare
FCC	Federal Communications Commission
FDD	frequency division duplex
FDM	frequency division multiplexing
FDMA	frequency division multiple access
FETs	field-effect transistors
FFHSS	fast frequency- hopping spread spectrum
FHSS	frequency-hopping spread spectrum
FLAR	forward-looking automotive radar
FMCW	frequency-modulated continous wave
FM	frequency modulated
FNBW	first-null beamwidth
FSK	frequency shift keying
GEO	geosynchronous orbit
GMSK	Gaussian minimum shift keying
GPS	global positioning system
GSM	global system for mobile communication
G/T	receiver antenna gain to system noise temperature ratio
HBTs	heterojunction bipolar transistors
HEMTs	high-electron-mobility transistors
HPA	high-power amplifier
HPBW	half-power beamwidth
HPF	high-pass filter
IF	intermediate frequency
IFM	instantaneous frequency measurement
IL	insertion loss
IMD	intermodulation distortion
IM	intermodulation
IMPATT	impact ionization avalanche transit time device
IM3	third-order intermodulation
IP3	third-order intercept point
I/Q	in-phase/quadrature-phase
IVHS	intelligent vehicle and highway system

JCT	Japanese cordless telephone
J/S	jammer-to-signal
LAN	local area network
LDMS	local multipoint distribution service
LEO	low earth orbit
LNA	low-noise amplifier
LO	local oscillator
LOS	line-of-sight
LPF	low-pass filter
LP	linearly polarized
MDS	minimum detectable signal
MEO	medium-altitude orbit
MESFETs	metal–semiconductor field-effect transistors
MIC	microwave integrated circuit
MLS	microwave landing system
MMIC	monolithic microwave integrated circuits
MSK	minimum shift keying
MTI	moving target indicator
NMT	Nordic mobile telephone
OQPSK	offset-keyed quadriphase shift keying
PAMELA	pricing and monitoring electronically of automobiles
PA	power amplifier
PAE	power added efficiency
PCM	pulse code modulation
PCN	personal communication networks
PCS	personal communication systems
PDC	personal digital cellular
PHS	personal handy phone system
PLL	phase-locked loops
PLO	phase-locked oscillators
PM	phase modulation
PN	pseudonoise
PRF	pulse repetition frequency
PSK	phase shift keying
8-PSK	8-phase shift keying
16-PSK	16-phase shift keying
QAM	quadrature amplitude modulation
QPSK	quadriphase shift keying
RCS	radar cross section
RF	radio frequency
RFID	radio frequency identification
RL	return loss
SAR	synthetic aperture radar
SAW	surface acoustic wave
SEP	spherical probable error

SFDR	spurious-free dynamic range
SFHSS	slow frequency- hopping spread spectrum
SLL	sidelobe levels
SMILER	short range microwave links for european roads
SNR	signal-noise ratio
SOJ	stand-off jammer
SPDT	single pole, double throw
SPST	single pole, single throw
SP3T	single pole, triple throw
SQPSK	staggered quadriphase shift keying
SS	spread spectrum
SS-CDMA	spread spectrum code division multiple access
SSJ	self-screening jammer
SSMI	special sensor microwave imager
STC	sensitivity time control
TACS	total access communication system
TDM	time division multiplexing
TDMA	time division multiple access
TE	transverse electric
TEM	transverse electromagnetic
TM	transverse magnetic
TOI	third-order intercept point
T/R	transmit/receive
TWTs	traveling-wave tubes
UHF	ultrahigh frequencies
VCO	voltage-controlled oscillator
VHF	very high frequency
VLF	very low frequency
VSWR	voltage standing-wave ratio
WLANs	wireless local-area networks

RF and Microwave
Wireless Systems

CHAPTER ONE

Introduction

1.1 BRIEF HISTORY OF RF AND MICROWAVE WIRELESS SYSTEMS

The wireless era was started by two European scientists, James Clerk Maxwell and Heinrich Rudolf Hertz. In 1864, Maxwell presented Maxwell's equations by unifying the works of Lorentz, Faraday, Ampere, and Gauss. He predicted the propagation of electromagnetic waves in free space at the speed of light. He postulated that light was an electromagnetic phenomenon of a particular wavelength and predicted that radiation would occur at other wavelengths as well. His theory was not well accepted until 20 years later, after Hertz validated the electromagnetic wave (wireless) propagation. Hertz demonstrated radio frequency (RF) generation, propagation, and reception in the laboratory. His radio system experiment consisted of an end-loaded dipole transmitter and a resonant square-loop antenna receiver operating at a wavelength of 4 m. For this work, Hertz is known as the father of radio, and frequency is described in units of hertz (Hz).

Hertz's work remained a laboratory curiosity for almost two decades, until a young Italian, Guglielmo Marconi, envisioned a method for transmitting and receiving information. Marconi commercialized the use of electromagnetic wave propagation for wireless communications and allowed the transfer of information from one continent to another without a physical connection. The telegraph became the means of fast communications. Distress signals from the *S.S. Titanic* made a great impression on the public regarding the usefulness of wireless communications. Marconi's wireless communications using the telegraph meant that a ship was no longer isolated in the open seas and could have continuous contact to report its positions. Marconi's efforts earned him the Nobel Prize in 1909.

In the early 1900s, most wireless transmission occurred at very long wavelengths. Transmitters consisted of Alexanderson alternators, Poulsen arcs, and spark gaps. Receivers used coherers, Fleming valves, and DeForest audions. With the advent of DeForest's triode vacuum tube in 1907, continuous waves (CW) replaced spark gaps, and more reliable frequency and power output were obtained for radio broadcasting at frequencies below 1.5 MHz. In the 1920s, the one-way broadcast was made to police cars in Detroit. Then the use of radio waves for wireless broadcasting, communications between mobile and land stations, public safety systems, maritime mobile services, and land transportation systems was drastically increased. During World War II, radio communications became indispensable for military use in battlefields and troop maneuvering.

World War II also created an urgent need for radar (standing for radio detection and ranging). The acronym radar has since become a common term describing the use of reflections from objects to detect and determine the distance to and relative speed of a target. A radar's resolution (i.e., the minimum object size that can be detected) is proportional to wavelength. Therefore, shorter wavelengths or higher frequencies (i.e., microwave frequencies and above) are required to detect smaller objects such as fighter aircraft.

Wireless communications using telegraphs, broadcasting, telephones, and point-to-point radio links were available before World War II. The widespread use of these communication methods was accelerated during and after the war. For long-distance wireless communications, relay systems or tropospheric scattering were used. In 1959, J. R. Pierce and R. Kompfner envisioned transoceanic communications by satellites. This opened an era of global communications using satellites. The satellite uses a broadband high-frequency system that can simultaneously support thousands of telephone users, tens or hundreds of TV channels, and many data links. The operating frequencies are in the gigahertz range. After 1980, cordless phones and

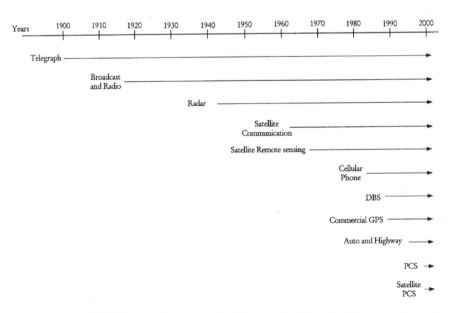

FIGURE 1.1 Summary of the history of wireless systems.

cellular phones became popular and have enjoyed very rapid growth in the past two decades. Today, personal communication systems (PCSs) operating at higher frequencies with wider bandwidths are emerging with a combination of various services such as voice mail, email, video, messaging, data, and computer on-line services. The direct link between satellites and personal communication systems can provide voice, video, or data communications anywhere in the world, even in the most remote regions of the globe.

In addition to communication and radar applications, wireless technologies have many other applications. In the 1990s, the use of wireless RF and microwave technologies for motor vehicle and highway applications has increased, especially in Europe and Japan. The direct broadcast satellite (DBS) systems have offered an alternative to cable television, and the end of the Cold War has made many military technologies available to civilian applications. The global positioning systems (GPSs), RF identification (RFID) systems, and remote sensing and surveillance systems have also found many commercial applications.

Figure 1.1 summarizes the history of these wireless systems.

1.2 FREQUENCY SPECTRUMS

Radio frequencies, microwaves, and millimeter waves occupy the region of the electromagnetic spectrum below 300 GHz. The microwave frequency spectrum is from 300 MHz to 30 GHz with a corresponding wavelength from 100 cm to 1 cm. Below the microwave spectrum is the RF spectrum and above is the millimeter-wave spectrum. Above the millimeter-wave spectrum are submillimeter-wave, infrared, and optical spectrums. Millimeter waves (30–300 GHz), which derive their name from the dimensions of the wavelengths (from 10 to 1 mm), can be classified as microwaves since millimeter-wave technology is quite similar to that of microwaves. Figure 1.2 shows the electromagnetic spectrum. For convenience, microwave and millimeter-wave spectrums are further divided into many frequency bands. Figure 1.2 shows some microwave bands, and Table 1.1 shows some millimeter-wave bands. The RF spectrum is not well defined. One can consider the frequency spectrum below 300 MHz as the RF spectrum. But frequently, literatures use the RF term up to 2 GHz or even higher.

The Federal Communications Commission (FCC) allocates frequency ranges and specifications for different applications in the United States, including televisions, radios, satellite communications, cellular phones, police radar, burglar alarms, and navigation beacons. The performance of each application is strongly affected by the atmospheric absorption. The absorption curves are shown in Fig. 1.3. For example, a secure local area network would be ideal at 60 GHz due to the high attenuation caused by the O_2 resonance.

As more applications spring up, overcrowding and interference at lower frequency bands pushes applications toward higher operating frequencies. Higher frequency operation has several advantages, including:

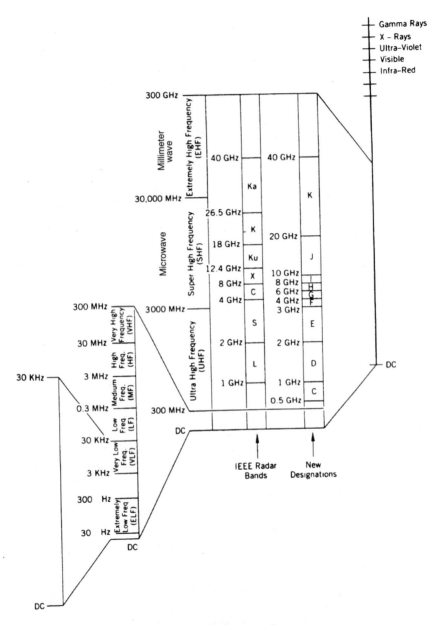

FIGURE 1.2   Electromagnetic spectrum.

1. Larger instantaneous bandwidth for greater transfer of information
2. Higher resolution for radar, bigger doppler shift for CW radar, and more detailed imaging and sensing
3. Reduced dimensions for antennas and other components
4. Less interference from nearby applications
5. Fast speed for digital system signal processing and data transmission
6. Less crowded spectrum
7. Difficulty in jamming (military applications)

TABLE 1.1 Millimeter-Wave Band Designation

Designation	Frequency Range (GHz)
Q-band	33–50
U-band	40–60
V-band	50–75
E-band	60–90
W-band	75–110
D-band	110–170
G-band	140–220
Y-band	220–325

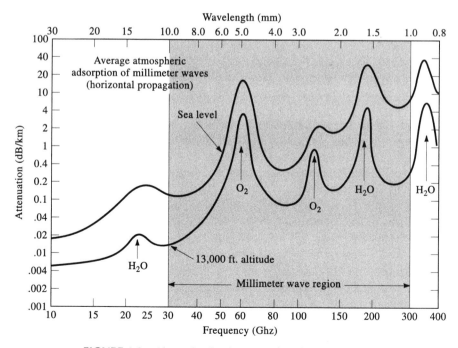

FIGURE 1.3 Absorption by the atmosphere in clear weather.

The use of higher frequency also has some disadvantages:

1. More expensive components
2. Higher atmospheric losses
3. Reliance on GaAs instead of Si technology
4. Higher component losses and lower output power from active devices
5. Less accurate design tools and less mature technologies

The electron mobility in GaAs is higher than that in silicon. Therefore, GaAs devices can operate at higher frequencies and speeds. Current silicon-based devices are commonly used up to 2 GHz. Above 4 GHz, GaAs devices are preferred for better performance. However, GaAs processing is more expensive, and the yield is lower than that of silicon.

1.3   WIRELESS APPLICATIONS

Two of the most historically important RF/microwave applications are communication systems and radar; but there are many others. Currently, the market is driven by the phenomenal growth of PCSs, although there is also an increased demand for satellite-based video, telephone, and data communication systems.

Radio waves and microwaves play an important role in modern life. Television signals are transmitted around the globe by satellites using microwaves. Airliners are guided from takeoff to landing by microwave radar and navigation systems. Telephone and data signals are transmitted using microwave relays. The military uses microwaves for surveillance, navigation, guidance and control, communications, and identification in their tanks, ships, and planes. Cellular telephones are everywhere.

The RF and microwave wireless technologies have many commercial and military applications. The major application areas include communications, radar, navigation, remote sensing, RF identification, broadcasting, automobiles and highways, sensors, surveillance, medical, and astronomy and space exploration. The details of these applications are listed below:

1. *Wireless Communications.* Space, long-distance, cordless phones, cellular telephones, mobile, PCSs, local-area networks (LANs), aircraft, marine, citizen's band (CB) radio, vehicle, satellite, global, etc.
2. *Radar.* Airborne, marine, vehicle, collision avoidance, weather, imaging, air defense, traffic control, police, intrusion detection, weapon guidance, surveillance, etc.
3. *Navigation.* Microwave landing system (MLS), GPS, beacon, terrain avoidance, imaging radar, collision avoidance, auto-pilot, aircraft, marine, vehicle, etc.

4. *Remote Sensing*. Earth monitoring, meteorology, pollution monitoring, forest, soil moisture, vegetation, agriculture, fisheries, mining, water, desert, ocean, land surface, clouds, precipitation, wind, flood, snow, iceberg, urban growth, aviation and marine traffic, surveillance, etc.

5. *RF Identification*. Security, antitheft, access control, product tracking, inventory control, keyless entry, animal tracking, toll collection, automatic checkout, asset management, etc.

6. *Broadcasting*. Amplitude- and frequency-modulated (AM, FM) radio, TV, DBS, universal radio system, etc.

7. *Automobiles and Highways*. Collision warning and avoidance, GPS, blind-spot radar, adaptive cruise control, autonavigation, road-to-vehicle communications, automobile communications, near-obstacle detection, radar speed sensors, vehicle RF identification, intelligent vehicle and highway system (IVHS), automated highway, automatic toll collection, traffic control, ground penetration radar, structure inspection, road guidance, range and speed detection, vehicle detection, etc.

8. *Sensors*. Moisture sensors, temperature sensors, robotics, buried-object detection, traffic monitoring, antitheft, intruder detection, industrial sensors, etc.

9. *Surveillance and Electronic Warfare*. Spy satellites, signal or radiation monitoring, troop movement, jamming, antijamming, police radar detectors, intruder detection, etc.

10. *Medical*. Magnetic resonance imaging, microwave imaging, patient monitoring, etc.

11. *Radio Astronomy and Space Exploration*. Radio telescopes, deep-space probes, space monitoring, etc.

12. *Wireless Power Transmission*. Space to space, space to ground, ground to space, ground to ground power transmission.

1.4   A SIMPLE SYSTEM EXAMPLE

A wireless system is composed of active and passive devices interconnected to perform a useful function. A simple example of a wireless radio system is shown in Fig. 1.4.

The transmitter operates as follows. The input baseband signal, which could be voice, video, or data, is assumed to be bandlimited to a frequency f_m. This signal is filtered to remove any components that may be beyond the channel's passband. The message signal is then mixed with a local oscillator (LO) signal to produce a modulated carrier in a process called up-conversion since it produces signals at frequencies $f_{LO} + f_m$ or $f_{LO} - f_m$ which are normally much higher than f_m. The modulated carrier can then be amplified and transmitted by the antenna.

When the signal arrives at the receiver, it is normally amplified by a low-noise amplifier (LNA). The LNA may be omitted from some systems when the received signal has enough power to be mixed directly, as may occur in short-distance

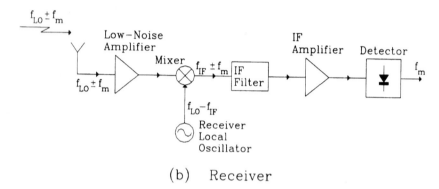

FIGURE 1.4   Block diagram of a simplified wireless radio system.

communication links. The mixer then produces a signal at a frequency $f_{IF} + f_m$ or $f_{IF} - f_m$ in a process called down-conversion since f_{IF} is chosen to be much lower than f_{LO}. The signal is filtered to remove any undesired harmonic and spurious products resulting from the mixing process and is amplified by an intermediate-frequency (IF) amplifier. The output of the amplifier goes to a detector stage where the baseband signal f_m, which contains the original message, is recovered.

To perform all of these functions, the microwave system relies on separate components that contribute specific functions to the overall system performance. Broadly speaking, microwave components can be classified as transmission lines, couplers, filters, resonators, signal control components, amplifiers, oscillators, mixers, detectors, and antennas.

1.5   ORGANIZATION OF THIS BOOK

This book is organized into 11 chapters. Chapter 2 reviews some fundamental principles of transmission lines and electromagnetic waves. Chapter 3 gives a brief overview of how antennas and antenna arrays work. Chapter 4 provides a discussion of system parameters for various components that form building blocks for a system. Chapters 5 and 6 are devoted to receiver and transmitter systems. Chapter 7 introduces the radar equation and the basic principles for pulse and CW radar systems. Chapter 8 describes various wireless communication systems, including radios, microwave links, satellite communications, and cellular phones. It also discusses the Friis transmission equation, space loss, and link budget calculation. A brief introduction to modulation techniques and multiple-access methods is given in Chapters 9 and 10. Finally, Chapter 11 describes other wireless applications, including navigation systems, automobile and highway applications, direct broadcast systems, remote sensing systems, RF identification systems, and surveillance systems.

CHAPTER TWO

Review of Waves and Transmission Lines

2.1 INTRODUCTION

At low RF, a wire or a line on a printed circuit board can be used to connect two electronic components. At higher frequencies, the current tends to concentrate on the surface of the wire due to the skin effect. The skin depth is a function of frequency and conductivity given by

$$\delta_s = \left(\frac{2}{\omega\mu\sigma}\right)^{1/2} \tag{2.1}$$

where $\omega = 2\pi f$ is the angular frequency, f is the frequency, μ is the permeability, and σ is the conductivity. For copper at a frequency of 10 GHz, $\sigma = 5.8 \times 10^7$ S/m and $\delta_s = 6.6 \times 10^{-5}$ cm, which is a very small distance. The field amplitude decays exponentially from its surface value according to e^{-z/δ_s}, as shown in Fig. 2.1. The field decays by an amount of e^{-1} in a distance of skin depth δ_s. When a wire is operating at low RF, the current is distributed uniformly inside the wire, as shown in Fig. 2.2. As the frequency is increased, the current will move to the surface of the wire. This will cause higher conductor losses and field radiation. To overcome this problem, shielded wires or field-confined lines are used at higher frequencies.

Many transmission lines and waveguides have been proposed and used in RF and microwave frequencies. Figure 2.3 shows the cross-sectional views of some of these structures. They can be classified into two categories: conventional and integrated circuits. A qualitative comparison of some of these structures is given in Table 2.1. Transmission lines and/or waveguides are extensively used in any system. They are used for interconnecting different components. They form the building blocks of

10

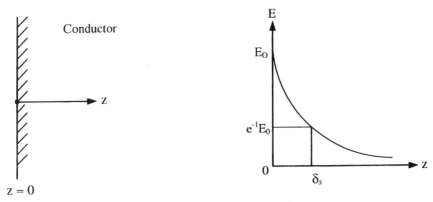

FIGURE 2.1 Fields inside the conductor.

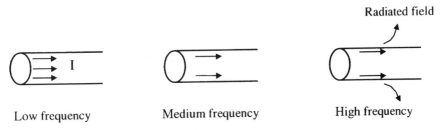

FIGURE 2.2 The currrent distribution within a wire operating at different frequencies.

many components and circuits. Examples are the matching networks for an amplifier and sections for a filter. They can be used for wired communications to connect a transmitter to a receiver (Cable TV is an example).

The choice of a suitable transmission medium for constructing microwave circuits, components, and subsystems is dictated by electrical and mechanical trade-offs. Electrical trade-offs involve such parameters as transmission line loss, dispersion, higher order modes, range of impedance levels, bandwidth, maximum operating frequency, and suitability for component and device implementation. Mechanical trade-offs include ease of fabrication, tolerance, reliability, flexibility, weight, and size. In many applications, cost is an important consideration.

This chapter will discuss the transmission line theory, reflection and transmission, S-parameters, and impedance matching techniques. The most commonly used transmission lines and waveguides such as coaxial cables, microstrip lines, and rectangular waveguides will be described.

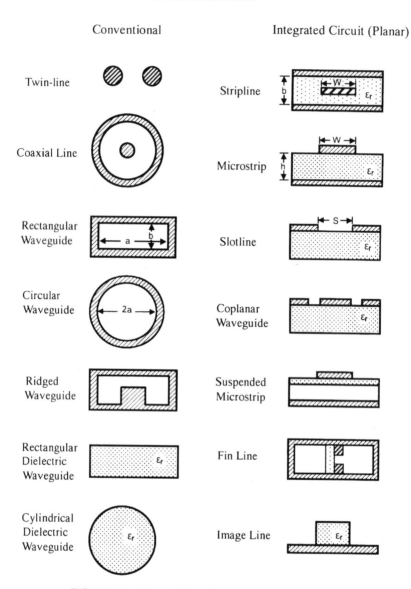

FIGURE 2.3   Transmission line and waveguide structures.

2.2   WAVE PROPAGATION

Waves can propagate in free space or in a transmission line or waveguide. Wave propagation in free space forms the basis for wireless applications. Maxwell predicted wave propagation in 1864 by the derivation of the wave equations. Hertz validated Maxwell's theory and demonstrated radio wave propagation in the

TABLE 2.1 Transmission Line and Waveguide Comparisons

Transmission Line	Useful Frequency Range (GHz)	Impedance Range (Ω)	Cross-Sectional Dimensions	Q-Factor	Power Rating	Active Device Mounting	Potential for Low-Cost Production
Rectangular waveguide	<300	100–500	Moderate to large	High	High	Easy	Poor
Coaxial line	<50	10–100	Moderate	Moderate	Moderate	Fair	Poor
Stripline	<10	10–100	Moderate	Low	Low	Fair	Good
Microstrip line	≤100	10–100	Small	Low	Low	Easy	Good
Suspended stripline	≤150	20–150	Small	Moderate	Low	Easy	Fair
Finline	≤150	20–400	Moderate	Moderate	Low	Easy	Fair
Slotline	≤60	60–200	Small	Low	Low	Fair	Good
Coplanar waveguide	≤60	40–150	Small	Low	Low	Fair	Good
Image guide	<300	30–30	Moderate	High	Low	Poor	Good
Dielectric line	<300	20–50	Moderate	High	Low	Poor	Fair

laboratory in 1886. This opened up an era of radio wave applications. For his work, Hertz is known as the father of radio, and his name is used as the frequency unit.

Let us consider the following four Maxwell equations:

$$\nabla \cdot \vec{E} = \frac{\rho}{\varepsilon} \quad \text{Gauss' law} \tag{2.2a}$$

$$\nabla \times \vec{E} = -\frac{\partial \vec{B}}{\partial t} \quad \text{Faraday's law} \tag{2.2b}$$

$$\nabla \times \vec{H} = \frac{\partial \vec{D}}{\partial t} + \vec{J} \quad \text{Ampere's law} \tag{2.2c}$$

$$\nabla \cdot \vec{B} = 0 \quad \text{flux law} \tag{2.2d}$$

where $\vec{E}$ and $\vec{B}$ are electric and magnetic fields, $\vec{D}$ is the electric displacement, $\vec{H}$ is the magnetic intensity, $\vec{J}$ is the conduction current density, ε is the permittivity, and ρ is the charge density. The term $\partial \vec{D} / \partial t$ is displacement current density, which was first added by Maxwell. This term is important in leading to the possibility of wave propagation. The last equation is for the continuity of flux.

We also have two constitutive relations:

$$\vec{D} = \varepsilon_0 \vec{E} + \vec{P} = \varepsilon \vec{E} \tag{2.3a}$$

$$\vec{B} = \mu_0 \vec{H} + \vec{M} = \mu \vec{H} \tag{2.3b}$$

where $\vec{P}$ and $\vec{M}$ are the electric and magnetic dipole moments, respectively, μ is the permeability, and ε is the permittivity. The relative dielectric constant of the medium and the relative permeability are given by

$$\varepsilon_r = \frac{\varepsilon}{\varepsilon_0} \tag{2.4a}$$

$$\mu_r = \frac{\mu}{\mu_0} \tag{2.4b}$$

where $\mu_0 = 4\pi \times 10^{-7}$ H/m is the permeability of vacuum and $\varepsilon_0 = 8.85 \times 10^{-12}$ F/m is the permittivity of vacuum.

With Eqs. (2.2) and (2.3), the wave equation can be derived for a source-free transmission line (or waveguide) or free space. For a source-free case, we have $\vec{J} = \rho = 0$, and Eq. (2.2) can be rewritten as

$$\nabla \cdot \vec{E} = 0 \tag{2.5a}$$

$$\nabla \times \vec{E} = -j\omega\mu \vec{H} \tag{2.5b}$$

$$\nabla \times \vec{H} = j\omega\varepsilon \vec{E} \tag{2.5c}$$

$$\nabla \cdot \vec{H} = 0 \tag{2.5d}$$

Here we assume that all fields vary as $e^{j\omega t}$ and $\partial/\partial t$ is replaced by $j\omega$.

The curl of Eq. (2.5b) gives

$$\nabla \times \nabla \times \vec{E} = -j\omega\mu\nabla \times \vec{H} \tag{2.6}$$

Using the vector identity $\nabla \times \nabla \times \vec{E} = \nabla(\nabla \cdot \vec{E}) - \nabla^2\vec{E}$ and substituting (2.5c) into Eq. (2.6), we have

$$\nabla(\nabla \cdot \vec{E}) - \nabla^2\vec{E} = -j\omega\mu(j\omega\varepsilon\vec{E}) = \omega^2\mu\varepsilon\vec{E} \tag{2.7}$$

Substituting Eq. (2.5a) into the above equation leads to

$$\nabla^2\vec{E} + \omega^2\mu\varepsilon\vec{E} = 0 \tag{2.8a}$$

or

$$\nabla^2\vec{E} + k^2\vec{E} = 0 \tag{2.8b}$$

where $k = \omega\sqrt{\mu\varepsilon}$ = propagation constant.

Similarly, one can derive

$$\nabla^2\vec{H} + \omega^2\mu\varepsilon\vec{H} = 0 \tag{2.9}$$

Equations (2.8) and (2.9) are referred to as the Helmholtz equations or wave equations. The constant k (or β) is called the wave number or propagation constant, which may be expressed as

$$k = \omega\sqrt{\mu\varepsilon} = \frac{2\pi}{\lambda} = 2\pi\frac{f}{v} = \frac{\omega}{v} \tag{2.10}$$

where λ is the wavelength and v is the wave velocity.

In free space or air-filled transmission lines, $\mu = \mu_0$ and $\varepsilon = \varepsilon_0$, we have $k = k_0 = \omega\sqrt{\mu_0\varepsilon_0}$, and $v = c = 1/\sqrt{\mu_0\varepsilon_0}$ = speed of light. Equations (2.8) and (2.9) can be solved in rectangular, cylindrical, or spherical coordinates. Antenna radiation in free space is an example of spherical coordinates. The solution in a wave propagating in the $\vec{r}$ direction:

$$\vec{E}(r, \theta, \phi) = \vec{E}(\theta, \phi)e^{-j\vec{k}\cdot\vec{r}} \tag{2.11}$$

The propagation in a rectangular waveguide is an example of rectangular coordinates with a wave propagating in the z direction:

$$\vec{E}(x, y, z) = \vec{E}(x, y)e^{-jkz} \tag{2.12}$$

Wave propagation in cylindrical coordinates can be found in the solution for a coaxial line or a circular waveguide with the field given by

$$\vec{E}(r, \phi, z) = \vec{E}(r, \phi)e^{-jkz} \tag{2.13}$$

From the above discussion, we can conclude that electromagnetic waves can propagate in free space or in a transmission line. The wave amplitude varies with time as a function of $e^{j\omega t}$. It also varies in the direction of propagation and in the transverse direction. The periodic variation in time as shown in Fig. 2.4 gives the

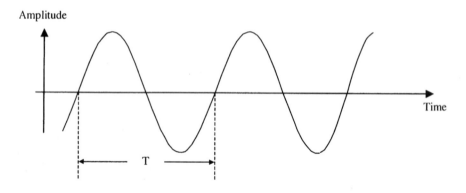

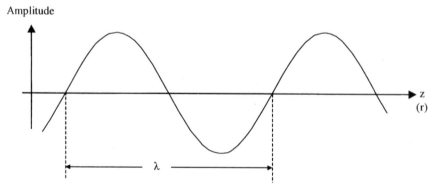

FIGURE 2.4   Wave variation in time and space domains.

frequency f, which is equal to $1/T$, where T is the period. The period length in the propagation direction gives the wavelength. The wave propagates at a speed as

$$v = f\lambda \tag{2.14}$$

Here, v equals the speed of light c if the propagation is in free space:

$$v = c = f\lambda_0 \tag{2.15}$$

λ_0 being the free-space wavelength.

2.3 TRANSMISSION LINE EQUATION

The transmission line equation can be derived from circuit theory. Suppose a transmission line is used to connect a source to a load, as shown in Fig. 2.5. At position x along the line, there exists a time-varying voltage $v(x, t)$ and current $i(x, t)$. For a small section between x and $x + \Delta x$, the equivalent circuit of this section Δx can be represented by the distributed elements of L, R, C, and G, which are the inductance, resistance, capacitance, and conductance per unit length. For a lossless line, $R = G = 0$. In most cases, R and G are small. This equivalent circuit can be easily understood by considering a coaxial line in Fig. 2.6. The parameters L and R are due to the length and conductor losses of the outer and inner conductors, whereas

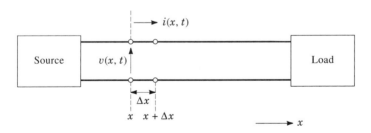

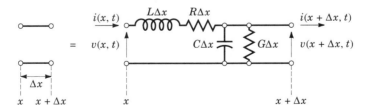

FIGURE 2.5 Transmission line equivalent circuit.

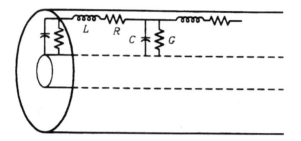

FIGURE 2.6 L, R, C for a coaxial line.

C and G are attributed to the separation and dielectric losses between the outer and inner conductors.

Applying Kirchhoff's current and voltage laws to the equivalent circuit shown in Fig. 2.5, we have

$$v(x + \Delta x, t) - v(x, t) = \Delta v(x, t) = -(R\,\Delta x)i(x, t)$$
$$- (L\,\Delta x)\frac{\partial i(x, t)}{\partial t} \tag{2.16}$$

$$i(x + \Delta x, t) - i(x, t) = \Delta i(x, t) = -(G\,\Delta x)v(x + \Delta x, t)$$
$$- (C\,\Delta x)\frac{\partial v(x + \Delta x, t)}{\partial t} \tag{2.17}$$

Dividing the above two equations by Δx and taking the limit as Δx approaches 0, we have the following equations:

$$\frac{\partial v(x, t)}{\partial x} = -Ri(x, t) - L\frac{\partial i(x, t)}{\partial t} \tag{2.18}$$

$$\frac{\partial i(x, t)}{\partial x} = -Gv(x, t) - C\frac{\partial v(x, t)}{\partial t} \tag{2.19}$$

Differentiating Eq. (2.18) with respect to x and Eq. (2.19) with respect to t gives

$$\frac{\partial^2 v(x, t)}{\partial x^2} = -R\frac{\partial i(x, t)}{\partial x} - L\frac{\partial^2 i(x, t)}{\partial x\,\partial t} \tag{2.20}$$

$$\frac{\partial^2 i(x, t)}{\partial t\,\partial x} = -G\frac{\partial v(x, t)}{\partial t} - C\frac{\partial^2 v(x, t)}{\partial t^2} \tag{2.21}$$

By substituting (2.19) and (2.21) into (2.20), one can eliminate $\partial i/\partial x$ and $\partial^2 i/(\partial x \, \partial t)$. If only the steady-state sinusoidally time-varying solution is desired, phasor notation can be used to simplify these equations [1, 2]. Here, v and i can be expressed as

$$v(x, t) = \text{Re}[V(x)e^{j\omega t}] \tag{2.22}$$

$$i(x, t) = \text{Re}[I(x)e^{j\omega t}] \tag{2.23}$$

where Re is the real part and ω is the angular frequency equal to $2\pi f$. A final equation can be written as

$$\frac{d^2 V(x)}{dx^2} - \gamma^2 V(x) = 0 \tag{2.24}$$

Note that Eq. (2.24) is a wave equation, and γ is the wave propagation constant given by

$$\gamma = [(R + j\omega L)(G + j\omega C)]^{1/2} = \alpha + j\beta \tag{2.25}$$

where α = attenuation constant in nepers per unit length

$\quad\;\; \beta$ = phase constant in radians per unit length.

The general solution to Eq. (2.24) is

$$V(x) = V_+ e^{-\gamma x} + V_- e^{\gamma x} \tag{2.26}$$

Equation (2.26) gives the solution for voltage along the transmission line. The voltage is the summation of a forward wave ($V_+ e^{-\gamma x}$) and a reflected wave ($V_- e^{\gamma x}$) propagating in the $+x$ and $-x$ directions, respectively.

The current $I(x)$ can be found from Eq. (2.18) in the frequency domain:

$$I(x) = I_+ e^{-\gamma x} - I_- e^{\gamma x} \tag{2.27}$$

where

$$I_+ = \frac{\gamma}{R + j\omega L} V_+, \qquad I_- = \frac{\gamma}{R + j\omega L} V_-$$

The characteristic impedance of the line is defined by

$$Z_0 = \frac{V_+}{I_+} = \frac{V_-}{I_-} = \frac{R + j\omega L}{\gamma} = \left(\frac{R + j\omega L}{G + j\omega C} \right)^{1/2} \tag{2.28}$$

For a lossless line, $R = G = 0$, we have

$$\gamma = j\beta = j\omega\sqrt{LC} \qquad (2.29a)$$

$$Z_0 = \sqrt{\frac{L}{C}} \qquad (2.29b)$$

$$\text{Phase velocity } v_p = \frac{\omega}{\beta} = f\lambda_g = \frac{1}{\sqrt{LC}} \qquad (2.29c)$$

where λ_g is the guided wavelength and β is the propagation constant.

2.4   REFLECTION, TRANSMISSION, AND IMPEDANCE FOR A TERMINATED TRANSMISSION LINE

The transmission line is used to connect two components. Figure 2.7 shows a transmission line with a length l and a characteristic impedance Z_0. If the line is lossless and terminated by a load Z_0, there is no reflection and the input impedance Z_{in} is always equal to Z_0 regardless of the length of the transmission line. If a load Z_L

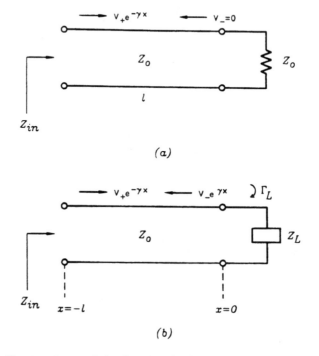

(a)

(b)

FIGURE 2.7   Terminated transmission line: (a) a load Z_0 is connected to a transmission line with a characteristic impedance Z_0; (b) a load Z_L is connected to a transmission line with a characteristic impedance Z_0.

(Z_L could be real or complex) is connected to the line as shown in Fig. 2.7b and $Z_L \neq Z_0$, there exists a reflected wave and the input impedance is no longer equal to Z_0. Instead, Z_{in} is a function of frequency (f), l, Z_L, and Z_0. Note that at low frequencies, $Z_{in} \approx Z_L$ regardless of l.

In the last section, the voltage along the line was given by

$$V(x) = V_+ e^{-\gamma x} + V_- e^{\gamma x} \tag{2.30}$$

A reflection coefficient along the line is defined as $\Gamma(x)$:

$$\Gamma(x) = \frac{\text{reflected } V(x)}{\text{incident } V(x)} = \frac{V_- e^{\gamma x}}{V_+ e^{-\gamma x}} = \frac{V_-}{V_+} e^{2\gamma x} \tag{2.31a}$$

where

$$\Gamma_L = \frac{V_-}{V_+} = \Gamma(0)$$

$$= \text{reflection coefficient at load} \tag{2.31b}$$

Substituting Γ_L into Eqs. (2.26) and (2.27), the impedance along the line is given by

$$Z(x) = \frac{V(x)}{I(x)} = Z_0 \frac{e^{-\gamma x} + \Gamma_L e^{\gamma x}}{e^{-\gamma x} - \Gamma_L e^{\gamma x}} \tag{2.32}$$

At $x = 0$, $Z(x) = Z_L$. Therefore,

$$Z_L = Z_0 \frac{1 + \Gamma_L}{1 - \Gamma_L} \tag{2.33}$$

$$\Gamma_L = |\Gamma_L| e^{j\phi} = \frac{Z_L - Z_0}{Z_L + Z_0} \tag{2.34}$$

To find the input impedance, we set $x = -l$ in $Z(x)$. This gives

$$Z_{in} = Z(-l) = Z_0 \frac{e^{\gamma l} + \Gamma_L e^{-\gamma l}}{e^{\gamma l} - \Gamma_L e^{-\gamma l}} \tag{2.35}$$

Substituting (2.34) into the above equation, we have

$$Z_{in} = Z_0 \frac{Z_L + Z_0 \tanh \gamma l}{Z_0 + Z_L \tanh \gamma l} \tag{2.36}$$

For the lossless case, $\gamma = j\beta$, (2.36) becomes

$$Z_{in} = Z_0 \frac{Z_L + jZ_0 \tan \beta l}{Z_0 + jZ_L \tan \beta l}$$

$$= Z_{in}(l, f, Z_L, Z_0) \tag{2.37}$$

Equation (2.37) is used to calculate the input impedance for a terminated lossless transmission line. It is interesting to note that, for low frequencies, $\beta l \approx 0$ and $Z_{in} \approx Z_L$.

The power transmitted and reflected can be calculated by the following:

$$\text{Incident power} = P_{in} = \frac{|V_+|^2}{Z_0} \tag{2.38}$$

$$\text{Reflected power} = P_r = \frac{|V_-|^2}{Z_0}$$

$$= \frac{|V_+|^2 |\Gamma_L|^2}{Z_0} = |\Gamma_L|^2 P_{in} \tag{2.39}$$

$$\text{Transmitted power} = P_t = P_{in} - P_r$$

$$= (1 - |\Gamma_L|^2) P_{in} \tag{2.40}$$

2.5   VOLTAGE STANDING-WAVE RATIO

For a transmission line with a matched load, there is no reflection, and the magnitude of the voltage along the line is equal to $|V_+|$. For a transmission line terminated with a load Z_L, a reflected wave exists, and the incident and reflected waves interfere to produce a standing-wave pattern along the line. The voltage at point x along the lossless line is given by

$$V(x) = V_+ e^{-j\beta x} + V_- e^{j\beta x}$$

$$= V_+ e^{-j\beta x}(1 + \Gamma_L e^{2j\beta x}) \tag{2.41}$$

Substituting $\Gamma_L = |\Gamma_L| e^{j\phi}$ into the above equation gives the magnitude of $V(x)$ as

$$|V(x)| = |V_+|[(1 + |\Gamma_L|)^2 - 4|\Gamma_L| \sin^2(\beta x + \tfrac{1}{2}\phi)]^{1/2} \tag{2.42}$$

Equation (2.42) shows that $|V(x)|$ oscillates between a maximum value of $|V_+|(1 + |\Gamma_L|)$ when $\sin(\beta x + \tfrac{1}{2}\phi) = 0$ (or $\beta x + \tfrac{1}{2}\phi = n\pi$) and a minimum value of $|V_+|(1 - |\Gamma_L|)$ when $\sin(\beta x + \tfrac{1}{2}\phi) = 1$ (or $\beta x + \tfrac{1}{2}\phi = m\pi - \tfrac{1}{2}\pi$). Figure 2.8 shows the pattern that repeats itself every $\tfrac{1}{2}\lambda_g$.

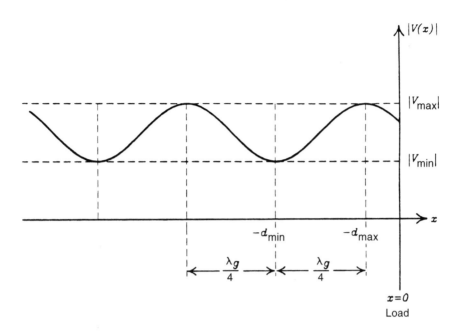

FIGURE 2.8   Pattern of voltage magnitude along line.

The first maximum voltage can be found by setting $x = -d_{max}$ and $n = 0$. We have

$$2\beta d_{max} = \phi \tag{2.43}$$

The first minimum voltage, found by setting $x = -d_{min}$ and $m = 0$, is given by

$$2\beta d_{min} = \phi + \pi \tag{2.44}$$

The voltage standing-wave ratio (VSWR) is defined as the ratio of the maximum voltage to the minimum voltage. From (2.42),

$$\text{VSWR} = \frac{|V_{max}|}{|V_{min}|} = \frac{1 + |\Gamma_L|}{1 - |\Gamma_L|} \tag{2.45}$$

If the VSWR is known, $|\Gamma_L|$ can be found by

$$|\Gamma_L| = \frac{\text{VSWR} - 1}{\text{VSWR} + 1} = |\Gamma(x)| \tag{2.46}$$

The VSWR is an important specification for all microwave components. For good matching, a low VSWR close to 1 is generally required over the operating frequency bandwidth. The VSWR can be measured by a VSWR meter together with a slotted line, a reflectometer, or a network analyzer. Figure 2.9 shows a nomograph of the VSWR. The return loss and power transmission are defined in the next section. Table 2.2 summarizes the formulas derived in the previous sections.

Example 2.1   Calculate the VSWR and input impedance for a transmission line connected to (a) a short and (b) an open load. Plot Z_{in} as a function of βl.

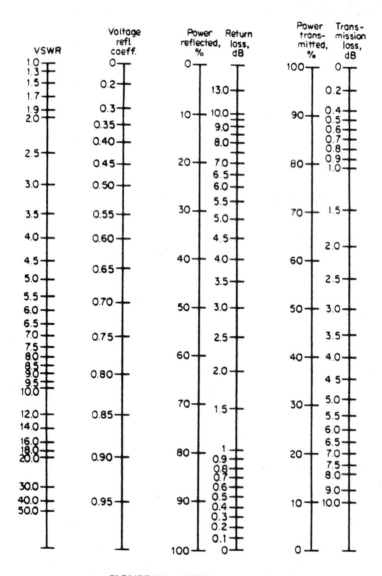

FIGURE 2.9   VSWR nomograph.

TABLE 2.2 **Formulas for Transmission Lines**

Quantity	General Line	Lossless Line
Propagation constant, $\gamma = \alpha + j\beta$	$\sqrt{(R + j\omega L)(G + j\omega C)}$	$j\omega\sqrt{LC}$
Phase constant, β	$\mathrm{Im}(\gamma)$	$\omega\sqrt{LC} = \dfrac{w}{v} = \dfrac{2\pi}{\lambda}$
Attenuation constant, α	$\mathrm{Re}(\gamma)$	0
Characteristic impedance, Z_0	$\sqrt{\dfrac{R + j\omega L}{G + j\omega C}}$	$\sqrt{\dfrac{L}{C}}$
Input impedance, Z_{in}	$Z_0 \dfrac{Z_L \cosh \gamma l + Z_0 \sinh \gamma l}{Z_0 \cosh \gamma l + Z_L \sinh \gamma l}$	$Z_0 \dfrac{Z_L \cos \beta l + jZ_0 \sin \beta l}{Z_0 \cos \beta l + jZ_L \sin \beta l}$
Impedance of shorted line	$Z_0 \tanh \gamma l$	$jZ_0 \tan \beta l$
Impedance of open line	$Z_0 \coth \gamma l$	$-jZ_0 \cot \beta l$
Impedance of quarter-wave line	$Z_0 \dfrac{Z_L \sinh \alpha l + Z_0 \cosh \alpha l}{Z_0 \sinh \alpha l + Z_L \cosh \alpha l}$	$\dfrac{Z_0^2}{Z_L}$
Impedance of half-wave line	$Z_0 \dfrac{Z_L \cosh \alpha l + Z_0 \sinh \alpha l}{Z_0 \cosh \alpha l + Z_L \sinh \alpha l}$	Z_L
Reflection coefficient, Γ_L	$\dfrac{Z_L - Z_0}{Z_L + Z_0}$	$\dfrac{Z_L - Z_0}{Z_L + Z_0}$
Voltage standing-wave ratio (VSWR)	$\dfrac{1 + \lvert\Gamma_L\rvert}{1 - \lvert\Gamma_L\rvert}$	$\dfrac{1 + \lvert\Gamma_L\rvert}{1 - \lvert\Gamma_L\rvert}$

Solution (a) A transmission line with a characteristic impedance Z_0 is connected to a short load Z_L as shown in Fig. 2.10. Here, $Z_L = 0$, and Γ_L is given by

$$\Gamma_L = \frac{Z_L - Z_0}{Z_L + Z_0} = -1 = 1e^{j180°} = \lvert\Gamma_L\rvert e^{j\phi}$$

Therefore, $\lvert\Gamma_L\rvert = 1$ and $\phi = 180°$. From Eq. (2.45),

$$\mathrm{VSWR} = \frac{1 + \lvert\Gamma_L\rvert}{1 - \lvert\Gamma_L\rvert} = \infty$$

$$\text{Reflected power} = \lvert\Gamma_L\rvert^2 P_{in} = P_{in}$$

$$\text{Transmitted power} = (1 - \lvert\Gamma_L\rvert^2)P_{in} = 0$$

The input impedance is calculated by Eq. (2.37),

$$Z_{in} = Z_0 \frac{Z_L + jZ_0 \tan \beta l}{Z_0 + jZ_L \tan \beta l}$$
$$= jZ_0 \tan \beta l = jX_{in}$$

The impedance Z_{in} is plotted as a function of βl or l, as shown in Fig. 2.10. It is interesting to note that any value of reactances can be obtained by varying l. For this

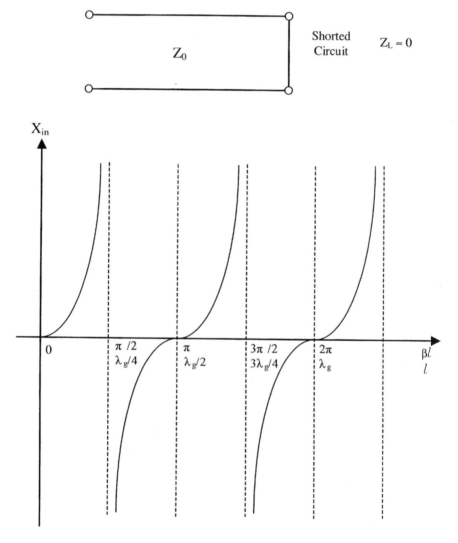

FIGURE 2.10   Transmission line connected to a shorted load.

reason, a short stub is useful for impedance tuning and impedance matching networks.

(b) For an open load, $Z_L = \infty$, and

$$\Gamma_L = \frac{Z_L - Z_0}{Z_L + Z_0} = 1 = 1e^{j0°} = |\Gamma_L|e^{j\phi}$$

Therefore, $|\Gamma_L| = 1$ and $\phi = 0°$. Again,

$$\text{VSWR} = \frac{1 + |\Gamma_L|}{1 - |\Gamma_L|} = \infty \quad \text{and} \quad P_r = P_{\text{in}}$$

The input impedance is given by

$$Z_{\text{in}} = Z_0 \frac{Z_L + jZ_0 \tan \beta l}{Z_0 + jZ_L \tan \beta l} = -jZ_0 \cot \beta l = jX_{\text{in}}$$

and is plotted as a function of βl or l as shown in Fig. 2.11. Again, any value of reactance can be obtained and the open stub can also be used as an impedance tuner. ■

2.6   DECIBELS, INSERTION LOSS, AND RETURN LOSS

The decibel (dB) is a dimensionless number that expresses the ratio of two power levels. Specifically,

$$\text{Power ratio in dB} = 10 \ \log_{10} \frac{P_2}{P_1} \tag{2.47}$$

where P_1 and P_2 are the two power levels being compared. If power level P_2 is higher than P_1, the decibel is positive and vice versa. Since $P = V^2/R$, the voltage definition of the decibel is given by

$$\text{Voltage ratio in dB} = 20 \ \log_{10} \frac{V_2}{V_1} \tag{2.48}$$

The decibel was originally named for Alexander Graham Bell. The unit was used as a measure of attenuation in telephone cable, that is, the ratio of the power of the signal emerging from one end of a cable to the power of the signal fed in at the other end. It so happened that one decibel almost equaled the attenuation of one mile of telephone cable.

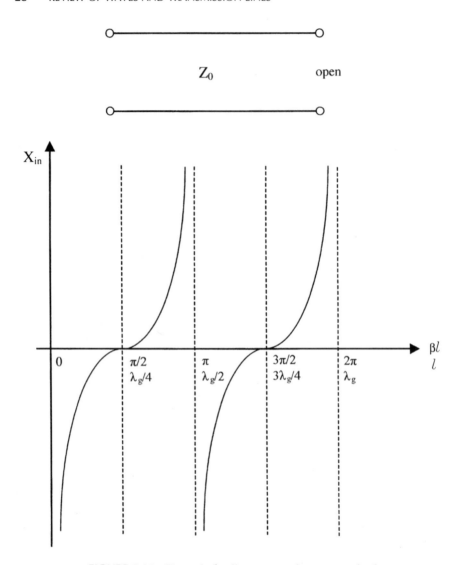

FIGURE 2.11 Transmission line connected to an open load.

2.6.1   Conversion from Power Ratios to Decibels and Vice Versa

One can convert any power ratio (P_2/P_1) to decibels, with any desired degree of accuracy, by dividing P_2 by P_1, finding the logarithm of the result, and multiplying it by 10.

From Eq. (2.47), we can find the power ratio in decibels given below:

Power Ratio	dB	Power Ratio	dB
1	0	1	0
1.26	1	0.794	-1
1.58	2	0.631	-2
2	3	0.501	-3
2.51	4	0.398	-4
3.16	5	0.316	-5
3.98	6	0.251	-6
5.01	7	0.2	-7
6.31	8	0.158	-8
7.94	9	0.126	-9
10	10	0.1	-10
100	20	0.01	-20
1000	30	0.001	-30
10^7	70	10^{-7}	-70

As one can see from these results, the use of decibels is very convenient to represent a very large or very small number. To convert from decibels to power ratios, the following equation can be used:

$$\text{Power ratio} = 10^{\text{dB}/10} \qquad (2.49)$$

2.6.2   Gain or Loss Representations

A common use of decibels is in expressing power gains and power losses in the circuits. Gain is the term for an increase in power level. As shown in Fig. 2.12, an amplifier is used to amplify an input signal with $P_{in} = 1\,\text{mW}$. The output signal is 200 mW. The amplifier has a gain given by

$$\text{Gain in ratio} = \frac{\text{output power}}{\text{input power}} = 200 \qquad (2.50a)$$

$$\text{Gain in db} = 10\,\log_{10}\frac{\text{output power}}{\text{input power}} = 23\,\text{dB} \qquad (2.50b)$$

Now consider an attenuator as shown in Fig. 2.13. The loss is the term of a decrease in power. The attenuator has a loss given by

$$\text{Loss in ratio} = \frac{\text{input power}}{\text{output power}} = 2 \qquad (2.51a)$$

$$\text{Loss in db} = 10\,\log_{10}\frac{\text{input power}}{\text{output power}} = 3\,\text{dB} \qquad (2.51b)$$

The above loss is called insertion loss. The insertion loss occurs in most circuit components, waveguides, and transmission lines. One can consider a 3-dB loss as a -3-dB gain.

For a cascaded circuit, one can add all gains (in decibels) together and subtract the losses (in decibels). Figure 2.14 shows an example. The total gain (or loss) is then

$$\text{Total gain} = 23 + 23 + 23 - 3 = 66 \text{ dB}$$

If one uses ratios, the total gain in ratio is

$$\text{Total gain} = 200 \times 200 \times 200 \times \tfrac{1}{2} = 4{,}000{,}000$$

In general, for any number of gains (in decibels) and losses (in decibels) in a cascaded circuit, the total gain or loss can be found by

$$G_T = (G_1 + G_2 + G_3 + \cdots) - (L_1 + L_2 + L_3 + \cdots) \tag{2.52}$$

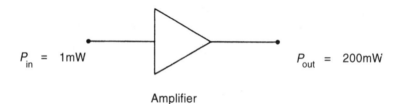

Amplifier

Gain = 200 or 23 dB

FIGURE 2.12   Amplifier circuit.

FIGURE 2.13   Attenuator circuit.

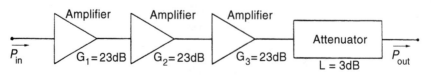

FIGURE 2.14   Cascaded circuit.

2.6.3   Decibels as Absolute Units

Decibels can be used to express values of power. All that is necessary is to establish some absolute unit of power as a reference. By relating a given value of power to this unit, the power can be expressed with decibels.

The often-used reference units are 1 mW and 1 W. If 1 milliwatt is used as a reference, dBm is expressed as decibels relative to 1 mW:

$$P \text{ (in dBm)} = 10 \ \log \ P \text{ (in mW)} \tag{2.53}$$

Therefore, the following results can be written:

$$1 \text{ mW} = 0 \text{ dBm}$$
$$10 \text{ mW} = 10 \text{ dBm}$$
$$1 \text{ W} = 30 \text{ dBm}$$
$$0.1 \text{ mW} = -10 \text{ dBm}$$
$$1 \times 10^{-7} \text{ mW} = -70 \text{ dBm}$$

If 1 W is used as a reference, dBW is expressed as decibels relative to 1 W. The conversion equation is given by

$$P \text{ (in dBW)} = 10 \ \log \ P \text{ (in W)} \tag{2.54}$$

From the above equation, we have

$$1 \text{ W} = 0 \text{ dBW} \qquad 10 \text{ W} = 10 \text{ dBW} \qquad 0.1 \text{ W} = -10 \text{ dBW}$$

Now for the system shown in Fig. 2.14, if the input power $P_{in} = 1$ mW $= 0$ dBm, the output power will be

$$\begin{aligned} P_{out} &= P_{in} + G_1 + G_2 + G_3 - L \\ &= 0 \text{ dBm} + 23 \text{ dB} + 23 \text{ dB} + 23 \text{ dB} - 3 \text{ dB} \\ &= 66 \text{ dBm, or } 3981 \text{ W} \end{aligned}$$

The above calculation is equivalent to the following equation using ratios:

$$\begin{aligned} P_{out} &= \frac{P_{in} G_1 G_2 G_3}{L} \\ &= \frac{1 \text{ mW} \times 200 \times 200 \times 200}{2} = 4000 \text{ W} \end{aligned}$$

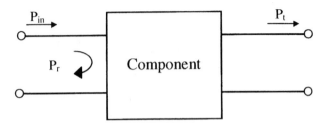

FIGURE 2.15 Two-port component.

2.6.4 Insertion Loss and Return Loss

Insertion loss, return loss, and VSWR are commonly used for component specification. As shown in Fig. 2.15, the insertion loss and return loss are defined as

$$\text{Insertion loss} = \text{IL} = 10 \ \log \ \frac{P_{\text{in}}}{P_t} \tag{2.55a}$$

$$\text{Return loss} = \text{RL} = 10 \ \log \ \frac{P_{\text{in}}}{P_r} \tag{2.55b}$$

Since $P_r = |\Gamma_L|^2 P_{\text{in}}$, Eq. (2.55b) becomes

$$\text{RL} = -20 \ \log \ |\Gamma_L| \tag{2.56}$$

The return loss indicates an input mismatch loss of a component. The insertion loss includes the input and output mismatch losses and other circuit losses (conductor loss, dielectric loss, and radiation loss).

Example 2.2 A coaxial three-way power divider (Fig. 2.16) has an input VSWR of 1.5 over a frequency range of 2.5–5.5 GHz. The insertion loss is 0.5 dB. What are the percentages of power reflection and transmission? What is the return loss in decibels?

Solution Since VSWR = 1.5, from Eq. (2.46)

$$|\Gamma_L| = \frac{\text{VSWR} - 1}{\text{VSWR} + 1} = \frac{0.5}{2.5} = 0.2 \qquad P_r = |\Gamma_L|^2 P_{\text{in}} = 0.04 P_{\text{in}}$$

The return loss is calculated by Eq. (2.55b) or (2.56):

$$\text{RL} = 10 \ \log \ \frac{P_{\text{in}}}{P_r} = -20 \ \log \ |\Gamma_L| = 13.98 \ \text{dB}$$

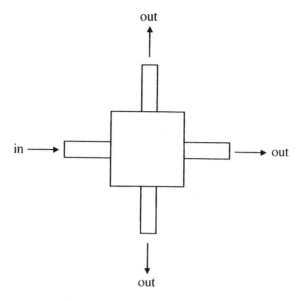

FIGURE 2.16 Three-way power divider.

The transmitted power is calculated from Eq. (2.55a):

$$\text{IL} = 10 \ \log \ \frac{P_{in}}{P_t} = 0.5 \text{ dB} \qquad P_t = 0.89 P_{in}$$

Assuming the input power is split into three output ports equally, each output port will transmit 29.7% of the input power. The input mismatch loss is 4% of the input power. Another 7% of power is lost due to the output mismatch and circuit losses.

◼

2.7 SMITH CHARTS

The Smith chart was invented by P. H. Smith of Bell Laboratories in 1939. It is a graphical representation of the impedance transformation property of a length of transmission line. Although the impedance and reflection information can be obtained from equations in the previous sections, the calculations normally involve complex numbers that can be complicated and time consuming. The use of the Smith chart avoids the tedious computation. It also provides a graphical representation on the impedance locus as a function of frequency.

Define a normalized impedance $\bar{Z}(x)$ as

$$\bar{Z}(x) = \frac{Z(x)}{Z_0} = \bar{R}(x) + j\bar{X}(x) \qquad (2.57)$$

The reflection coefficient $\Gamma(x)$ is given by

$$\Gamma(x) = \Gamma_r(x) + j\Gamma_i(x) \tag{2.58}$$

From Eqs. (2.31a) and (2.32), we have

$$\bar{Z}(x) = \frac{Z(x)}{Z_0} = \frac{1 + \Gamma(x)}{1 - \Gamma(x)} \tag{2.59}$$

Therefore

$$\bar{R}(x) + j\bar{X}(x) = \frac{1 + \Gamma_r + j\Gamma_i}{1 - \Gamma_r - j\Gamma_i} \tag{2.60}$$

By multiplying both numerator and denominator by $1 - \Gamma_r + j\Gamma_i$, two equations are generated:

$$\left(\Gamma_r - \frac{\bar{R}}{1 + \bar{R}}\right)^2 + \Gamma_i^2 = \left(\frac{1}{1 + \bar{R}}\right)^2 \tag{2.61a}$$

$$(\Gamma_r - 1)^2 + \left(\Gamma_i - \frac{1}{\bar{X}}\right)^2 = \left(\frac{1}{\bar{X}}\right)^2 \tag{2.61b}$$

In the $\Gamma_r - \Gamma_i$ coordinate system, Eq. (2.61a) represents circles centered at $(\bar{R}/(1 + \bar{R}),\ 0)$ with a radii of $1/(1 + \bar{R})$. These are called constant $\bar{R}$ circles. Equation (2.61b) represents circles centered at $(1,\ 1/\bar{X})$ with radii of $1/\bar{X}$. They are called constant $\bar{X}$ circles. Figure 2.17 shows these circles in the $\Gamma_r - \Gamma_i$ plane. The plot of these circles is called the Smith chart. On the Smith chart, a constant $|\Gamma|$ is a circle centered at $(0, 0)$ with a radius of $|\Gamma|$. Hence, motion along a lossless transmission line gives a circular path on the Smith chart. From Eq. (2.31), we know for a lossless line that

$$\Gamma(x) = \Gamma_L e^{2\gamma x} = \Gamma_L e^{j2\beta x} \tag{2.62}$$

Hence, given $\bar{Z}_L$, we can find $\bar{Z}_{\text{in}}$ at a distance $-l$ from the load by proceeding at an angle $2\beta l$ in a clockwise direction. Thus, $\bar{Z}_{\text{in}}$ can be found graphically.

The Smith chart has the following features:

1. Impedance or admittance values read from the chart are normalized values.
2. Moving away from the load (i.e., toward the generator) corresponds to moving in a clockwise direction.
3. A complete revolution around the chart is made by moving a distance $l = \frac{1}{2}\lambda_g$ along the transmission line.
4. The same chart can be used for reading admittance.

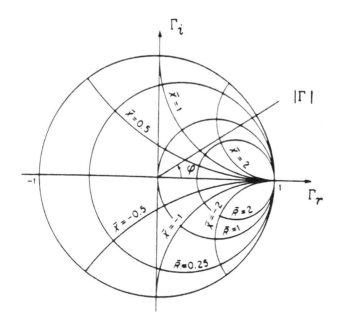

FIGURE 2.17 Constant $\bar{R}$ and $\bar{X}$ circles in the reflection-coefficient plane.

5. The center of the chart corresponds to the impedance-matched condition since $\Gamma(x) = 0$.

6. A circle centered at the origin is a constant $|\Gamma(x)|$ circle.

7. Moving along the lossless transmission line is equivalent to moving along the constant $|\Gamma(x)|$ circle.

8. For impedance reading, the point $(\Gamma_r = 1, \Gamma_i = 0)$ corresponds to an open circuit. For admittance reading, the same point corresponds to a short circuit.

9. The impedance at a distance $\frac{1}{4}\lambda_g$ from $\overline{Z}_L$ is equal to $\overline{Y}_L$.

10. The VSWR can be found by reading $\bar{R}$ at the intersection of the constant $|\Gamma|$ circle with the real axis.

The Smith chart can be used to find (1) Γ_L from Z_L and vice-versa; (2) $\overline{Z}_{in}$ from $\overline{Z}_L$ and vice-versa; (3) Z from Y and vice-versa; (4) the VSWR; and (5) d_{min} and d_{max}. The Smith chart is also useful for impedance matching, amplifier design, oscillator design, and passive component design.

The Smith charts shown in Fig. 2.17 are called Z-charts or Y-charts. One can read the normalized impedance or admittance directly from these charts. If we rotate the Y-chart by 180°, we have a rotated Y-chart, as shown in Fig. 2.18b. The combination of the Z-chart and the rotated Y-chart is called a Z–Y chart, as shown in Fig. 2.18c. On the Z–Y chart, for any point A, one can read $\overline{Z}_A$ from the Z-chart and $\overline{Y}_A$ from the rotated Y-chart.

Therefore, this chart avoids the necessity of moving $\bar{Z}$ by $\frac{1}{4}\lambda_g$ (i.e., 180°) to find $\bar{Y}$. The Z–Y chart is useful for impedance matching using lumped elements.

Example 2.3 A load of $100 + j50 \ \Omega$ is connected to a 50-Ω transmission line. Use both equations and the Smith chart to find the following: (a) Γ_L; (b) Z_{in} at $0.2\lambda_g$ away from the load; (c) Y_L; (d) the VSWR; and (e) d_{max} and d_{min}.

Solution

(a)

$$Z_L = 100 + j50$$

$$\bar{Z}_L = \frac{Z_L}{Z_0} = 2 + j$$

$$\Gamma_L = \frac{Z_L - Z_0}{Z_L + Z_0} = \frac{\bar{Z}_L - 1}{\bar{Z}_L + 1} = \frac{1 + j}{3 + j} = 0.4 + 0.2j$$

$$= 0.447 \angle 27°$$

From the Smith chart shown in Fig. 2.19, $\bar{Z}_L$ is located by finding the intersection of two circles with $\bar{R} = 2$ and $\bar{X} = 1$. The distance from the center to $\bar{Z}_L$ is 0.44, and the angle can be read directly from the scale as 27°.

(b)

$$Z_{\text{in}} = Z_0 \frac{Z_L + jZ_0 \tan \beta l}{Z_0 + jZ_L \tan \beta l}$$

$$l = 0.2\lambda_g, \qquad \beta l = \frac{2\pi}{\lambda_g} \times 0.2\lambda_g = 0.4\pi$$

$$\tan \beta l = 3.08$$

$$\bar{Z}_{\text{in}} = \frac{2 + j + j3.08}{1 + j(2 + j) \times 3.08} = 0.496 - j0.492$$

From the Smith chart, we move $\bar{Z}_L$ on a constant $|\Gamma|$ circle toward a generator by a distance $0.2\lambda_g$ to a point $\bar{Z}_{\text{in}}$. The value of $\bar{Z}_{\text{in}}$ can be read from the Smith chart as $0.5 - j0.5$, since $\bar{Z}_{\text{in}}$ is located at the intersection of two circles with $\bar{R} = 0.5$ and $\bar{X} = -0.5$. The distance of movement $(0.2\lambda_g)$ can be read from the outermost scale in the chart.

(c)

$$\bar{Z}_L = 2 + j$$

$$\bar{Y}_L = \frac{1}{\bar{Z}_L} = \frac{1}{2 + j} = 0.4 - j0.2$$

$$Y_L = \bar{Y}_L Y_0 = \frac{\bar{Y}_L}{Z_0} = \frac{0.4 - j0.2}{50} = 0.008 - j0.004$$

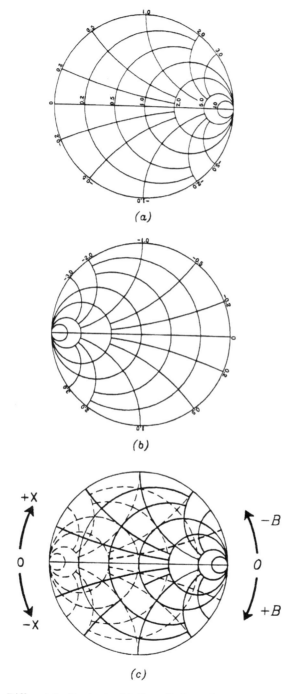

FIGURE 2.18 Different Smith charts: (a) Z- or Y-chart; (b) rotated Y-chart; and (c) Z–Y chart.

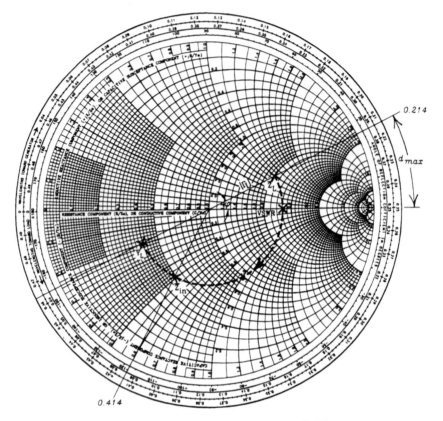

FIGURE 2.19   Smith chart for Example 2.3.

From the Smith chart, $\bar{Y}_L$ is found by moving $\bar{Z}_L$ a distance $0.25\lambda_g$ toward the generator. Therefore, $\bar{Y}_L$ is located opposite $\bar{Z}_L$ through the center. This can be proved by the following:

$$\bar{Z}_{\text{in}}\left(l = \frac{\lambda_g}{4}\right) = \frac{Z_L + jZ_0 \tan \beta l}{Z_0 + jZ_L \tan \beta l}$$

$$= \frac{Z_0}{Z_L} = \frac{1}{\bar{Z}_L} = \bar{Y}_L$$

(d)

$$\Gamma_L = 0.447\angle 27°$$

$$|\Gamma_L| = 0.447$$

$$\text{VSWR} = \frac{1 + |\Gamma_L|}{1 - |\Gamma_L|} = \frac{1 + 0.447}{1 - 0.447} = 2.62$$

The VSWR can be found from the reading of $\bar{R}$ at the constant $|\Gamma|$ circle and the real axis.

(e)   From Eq. (2.43), we have

$$d_{max} = \frac{\phi}{2\beta} = \frac{\phi}{4\pi}\lambda_g$$

Now

$$\phi = 27° \qquad d_{max} = 0.0375\lambda_g$$

From the Smith chart, d_{max} is the distance measured by moving from $\bar{Z}_L$ through a constant $|\Gamma|$ circle to the real axis. This can be proved by using Eq. (2.42). The first d_{max} occurs when $\beta x + \frac{1}{2}\phi = 0$ or $2\beta x + \phi = 0$. Since $\Gamma(x) = \Gamma_L e^{2j\beta x} = |\Gamma_L|e^{j(2\beta x + \phi)}$, the first d_{max} occurs when $\Gamma(x)$ has zero phase (i.e., on the real axis). The parameter d_{min} is just $\frac{1}{4}\lambda_g$ away from d_{max}.   ∎

2.8   S-PARAMETERS

Scattering parameters are widely used in RF and microwave frequencies for component modeling, component specifications, and circuit design. S-parameters can be measured by network analyzers and can be directly related to $ABCD$, Z-, and Y-parameters used for circuit analysis [1, 2]. For a general N-port network as shown in Fig. 2.20, the S-matrix is given in the following equations:

$$\begin{bmatrix} b_1 \\ b_2 \\ b_3 \\ \vdots \\ b_N \end{bmatrix} = \begin{bmatrix} S_{11} & S_{12} & S_{13} & \cdots & S_{1N} \\ S_{21} & S_{22} & S_{23} & \cdots & S_{2N} \\ \vdots & \vdots & \vdots & & \vdots \\ S_{N1} & S_{N2} & S_{N3} & \cdots & S_{NN} \end{bmatrix} \begin{bmatrix} a_1 \\ a_2 \\ a_3 \\ \vdots \\ a_N \end{bmatrix} \qquad (2.63)$$

or

$$[b] = [S][a] \qquad (2.64)$$

where $a_1, a_2, \ldots, a_N$ are the incident-wave voltages at ports $1, 2, \ldots, N$, respectively and $b_1, b_2, \ldots, b_N$ are the reflected-wave voltages at these ports. S-parameters are complex variables relating the reflected wave to the incident wave.

The S-parameters have some interesting properties [1, 2]:

1. For any matched port i, $S_{ii} = 0$.
2. For a reciprocal network, $S_{nm} = S_{mn}$.
3. For a passive circuit, $|S_{mn}| \leq 1$.

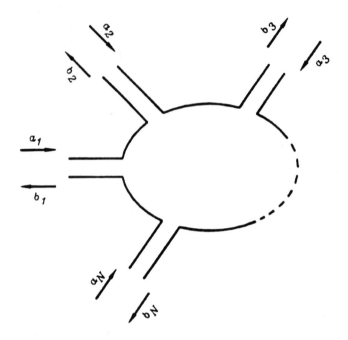

FIGURE 2.20   An N-port network.

4. For a lossless and reciprocal network, one has, for the ith port,

$$\sum_{n=1}^{N} |S_{ni}|^2 = \sum_{n=1}^{N} S_{ni}S_{ni}^* = 1 \tag{2.65}$$

or

$$|S_{1i}|^2 + |S_{2i}|^2 + |S_{3i}|^2 + \cdots + |S_{ii}|^2 + \cdots + |S_{Ni}|^2 = 1 \tag{2.66}$$

Equation (2.65) states that the product of any column of the S-matrix with the conjugate of this column is equal to 1. Equation (2.65) is due to the power conservation of a lossless network. In Eq. (2.65), the total power incident at the ith port is normalized and becomes 1, which is equal to the power reflected at the ith port plus the power transmitted into all other ports.

A two-port network is the most common circuit. For a two-port network, as shown in Fig. 2.21, the S-parameters are given by

$$b_1 = S_{11}a_1 + S_{12}a_2 \tag{2.67a}$$

$$b_2 = S_{21}a_1 + S_{22}a_2 \tag{2.67b}$$

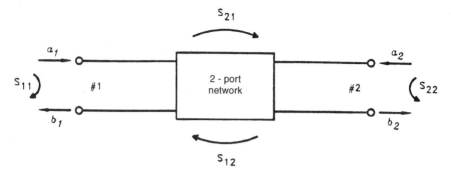

FIGURE 2.21 Two-port network and its S-parameters, which can be measured by using a network analyzer.

The S-parameters can be defined in the following:

$$S_{11} = \frac{b_1}{a_1}\bigg|_{a_2=0} = \Gamma_1 = \text{reflection coefficient at port 1 with } a_2 = 0$$

$$S_{21} = \frac{b_2}{a_1}\bigg|_{a_2=0} = T_{21} = \text{transmission coefficient from port 1 to 2 with } a_2 = 0$$

$$S_{22} = \frac{b_2}{a_2}\bigg|_{a_1=0} = \Gamma_2 = \text{reflection coefficient at port 2 with } a_1 = 0$$

$$S_{12} = \frac{b_1}{a_2}\bigg|_{a_1=0} = T_{12} = \text{transmission coefficient from port 2 to port 1 with } a_1 = 0$$

The return loss can be found by

$$RL = 20 \ \log\left|\frac{a_1}{b_1}\right| = 20 \ \log\left|\frac{1}{S_{11}}\right| \quad \text{in dB} \tag{2.68}$$

The attenuation or insertion loss is given by

$$IL = \alpha = 20 \ \log\left|\frac{a_1}{b_2}\right| = 20 \ \log\left|\frac{1}{S_{21}}\right| \quad \text{in dB} \tag{2.69}$$

The phase shift of the network is $\phi = $ phase of S_{21}.

2.9 COAXIAL LINES

Open-wire lines have high radiation losses and therefore are not used for high-frequency signal transmission. Shielded structures such as coaxial lines are commonly used. Cable TV and video transmission cables are coaxial lines. Coaxial lines are frequently used for interconnection, signal transmission, and measurements.

A coaxial line normally operates in the transverse electromagnetic (TEM) mode (i.e., no axial electric and magnetic field components). It has no cutoff frequency and can be used from direct current (DC) to microwave or millimeter-wave frequencies. It can be made flexible. Presently, coaxial lines and connectors can be operated up to 50 GHz. The upper frequency limit is set by the excessive losses and the excitation of circular waveguide transverse electric (TE) and transverse magnetic (TM) modes. However, improved manufacturing of small coaxial lines will push the operating frequency even higher.

Since the coaxial line is operated in the TEM mode, the electric and magnetic fields can be found in the static case. Figure 2.22 shows a coaxial transmission line. The electric potential of the coaxial line can be found by solving the Laplace equation in the cylindrical coordinates.

The Laplace equation for electric potential is

$$\nabla^2 \Phi = 0 \qquad (2.70)$$

Solving this equation by meeting the boundary conditions at the inner and outer conductors, we have [1, 2]

$$\Phi(r) = V_0 \frac{\ln(r/r_o)}{\ln(r_i/r_o)} \qquad (2.71)$$

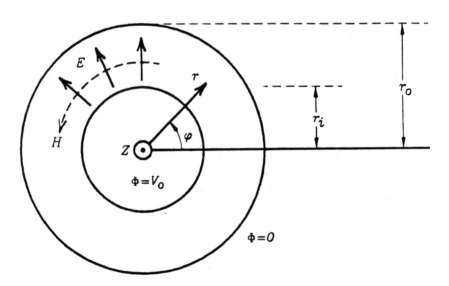

FIGURE 2.22   Coaxial line with inner and outer conductor radii r_i and r_o.

The electric field for a wave propagating in the positive z direction can be written as

$$\vec{E}(r) = -\nabla\Phi \, e^{-j\beta z} = \frac{V_o}{\ln(r_o/r_i)} \frac{1}{r} e^{-j\beta z} \hat{\underline{r}} \tag{2.72}$$

The magnetic field can be found from the Maxwell's equation

$$\vec{H} = -\frac{1}{j\omega\mu}\nabla \times \vec{E} = \frac{V_o}{\eta \ln(r_o/r_i)} \frac{1}{r} e^{-j\beta z} \hat{\underline{\phi}} \tag{2.73}$$

where $\eta = \sqrt{\mu/\varepsilon}$ is the wave impedance; $\hat{\underline{r}}$ and $\hat{\underline{\phi}}$ are unit vectors.

The characteristic impedance can be found from the current–voltage relation [1, 2]

$$Z_0 = \frac{V}{I} = \frac{\sqrt{\mu/\varepsilon}}{2\pi} \ln \frac{r_o}{r_i} \tag{2.74}$$

For an air-filled line, $\sqrt{\mu/\varepsilon} = \sqrt{\mu_0/\varepsilon_0} = 377\,\Omega$, and (2.74) becomes

$$Z_0 = 60 \ln \frac{r_o}{r_i} \quad (\Omega) \tag{2.75}$$

The above equations are useful for characteristic impedance calculation.

Commonly used coaxial lines have a characteristic impedance of $50\,\Omega$. The outer radii are 7, 3.5, and 2.4 mm operating up to frequencies of 18, 26, and 50 GHz, respectively. Some low-frequency coaxial lines have a characteristic impedance of $75\,\Omega$.

2.10 MICROSTRIP LINES

The microstrip line is the most commonly used microwave integrated circuit (MIC) transmission line. It has many advantages such as low cost, small size, absence of critical matching and cutoff frequency, ease of active device integration, use of photolithographic method for circuit production, good repeatability and reproducibility, ease of mass production, and compatibility with monolithic circuits. Monolithic microwave integrated circuits (MMICs) are microstrip circuits fabricated on GaAs or silicon substrates with both active devices and passive circuits on the same chip.

Compared to the rectangular waveguide and coaxial line circuits, microstrip lines disadvantages include higher loss, lower power-handling capability, and greater temperature instability. Synthesis and analysis formulas are well documented, and many discontinuities have been characterized. Commercial programs are available in order to use these models for circuit prediction and optimization.

Figure 2.23 shows the cross-section of a microstrip transmission line. A conductor strip with width w is etched on top of a substrate with thickness h. Two types of substrates are generally used: soft and hard. Soft substrate are flexible and

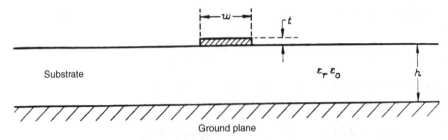

FIGURE 2.23 Microstrip transmission line.

cheap and can be easily fabricated. However, they have a higher thermal expansion coefficient. Some typical soft substrates are RT Duroid 5870 ($\varepsilon_r = 2.3$), RT Duroid 5880 ($\varepsilon_r = 2.2$), and RT Duroid 6010.5 ($\varepsilon_r = 10.5$). Hard substrates have better reliability and lower thermal expansion coefficients but are more expensive and inflexible. Typical hard substrates are quartz ($\varepsilon_r = 3.8$), alumina ($\varepsilon_r = 9.7$), sapphire ($\varepsilon_r = 11.7$), and GaAs ($\varepsilon_r = 12.3$). Table 2.3 provides the key design data for some common substrate materials [3].

TABLE 2.3 Properties of Microwave Dielectric Substrates

Material	Relative Dielectric Constant	Loss Tangent at 10 GHz (tan δ)	Thermal Conductivity, K (W/cm °C)	Dielectric Strength (kV/cm)
Sapphire	11.7	10^{-4}	0.4	4×10^3
Alumina	9.7	2×10^{-4}	0.3	4×10^3
Quartz (fused)	3.8	10^{-4}	0.01	10×10^3
Polystyrene	2.53	4.7×10^{-4}	0.0015	280
Beryllium oxide (BeO)	6.6	10^{-4}	2.5	—
GaAs ($\rho = 10^7$ Ω-cm)	12.3	16×10^{-4}	0.3	350
Si ($\rho = 10^3$ Ω-cm)	11.7	50×10^{-4}	0.9	300
3M 250 type GX	2.5	19×10^{-4}	0.0026	200
Keene Dl-clad 527	2.5	19×10^{-4}	0.0026	200
Rogers 5870	2.35	12×10^{-4}	0.0026	200
3M Cu-clad 233	2.33	12×10^{-4}	0.0026	200
Keene Dl-clad 870	2.33	12×10^{-4}	0.0026	200
Rogers 5880	2.20	9×10^{-4}	0.0026	200
3M Cu-clad 217	2.17	9×10^{-4}	0.0026	200
Keene Dl-clad 880	2.20	9×10^{-4}	0.0026	200
Rogers 6010	10.5	15×10^{-4}	0.004	160
3M epsilam 10	10.2	15×10^{-4}	0.004	160
Keene Dl-clad 810	10.2	15×10^{-4}	0.004	160
Air	1.0	0	0.00024	30

Source: From [3].

The most important parameters in microstrip circuit design are w, h and ε_r. The effects of strip thickness t and conductivity σ are secondary. Since the structure is not uniform, it supports the quasi-TEM mode. For a symmetrical structure such as a coaxial line or stripline, the TEM mode is supported. Analyses for microstrip lines can be complicated, and they are generally divided into two approaches. (1) Static or quasi-TEM analysis assumes a frequency equal to zero and solves the Laplace equation ($\nabla^2\Phi = 0$). The characteristic impedance is independent of frequency. The results are accurate only for low microwave frequencies. (2) Full-wave analysis solves $\nabla^2\Phi + k^2\Phi = 0$; therefore, the impedance is frequency dependent. The analysis provides good accuracy at millimeter-wave frequencies.

In the quasi-static analysis, the transmission line characteristics are calculated from the values of two capacitances, C and C_a, which are the unit-length capacitance with and without the dielectric substrate [2]. The characteristic impedance, phase velocity, guide wavelength, and effective dielectric constant are given as [2]

$$Z_0 = \frac{1}{c}\frac{1}{\sqrt{CC_a}} \tag{2.76}$$

$$v_p = c\sqrt{\frac{C_a}{C}} = \frac{c}{\sqrt{\varepsilon_{\text{eff}}}} \tag{2.77}$$

$$\varepsilon_{\text{eff}} = \frac{C}{C_a} = \left(\frac{\lambda_0}{\lambda_g}\right)^2 \tag{2.78}$$

$$\lambda_g = \frac{\lambda_0}{\sqrt{\varepsilon_{\text{eff}}}} \tag{2.79}$$

where c is the speed of light and λ_0 is the free-space wavelength. The above equations can be derived from Eq. (2.29) in the transmission line equation.

2.10.1 Analysis Formulas

The capacitances C and C_a can be found by solving $\nabla^2\Phi = 0$ for the cases with and without the substrate material. Many methods can be used to solve the Laplace equation meeting the boundary conditions. Some examples are the conformal mapping method [4], finite-difference method [5, 6], and variational method [7, 8]. Computer programs are available for these calculations. In practice, closed-form equations obtained from curve fitting are very handy for quick calculation. Some examples of design equations to find Z_0 and ε_{eff} given w/h and ε_r are [3]

$$\varepsilon_{\text{eff}} = \frac{\varepsilon_r + 1}{2} + \frac{\varepsilon_r - 1}{2}\left[\left(1 + \frac{12h}{w}\right)^{-1/2} + 0.04\left(1 - \frac{w}{h}\right)^2\right] \tag{2.80}$$

$$Z_0 = 60(\varepsilon_{\text{eff}})^{-1/2}\ln\left(\frac{8h}{w} + \frac{0.25w}{h}\right) \ \Omega \tag{2.81}$$

for $w/h \leq 1$ and

$$\varepsilon_{\text{eff}} = \frac{\varepsilon_r + 1}{2} + \frac{\varepsilon_r - 1}{2}\left(1 + \frac{12h}{w}\right)^{-1/2} \tag{2.82}$$

$$Z_0 = \frac{[120\pi(\varepsilon_{\text{eff}})^{-1/2}]}{(w/h) + 1.393 + 0.667 \; \ln(1.444 + w/h)} \quad \Omega \tag{2.83}$$

for $w/h > 1$.

2.10.2 Synthesis Formulas

Synthesis formulas are available for finding w/h and ε_{eff} if Z_0 and ε_r are known [9].
For narrow strips (i.e., when $Z_0 > 44 - 2\varepsilon_r$ ohms),

$$\frac{w}{h} = \left(\frac{\exp H'}{8} - \frac{1}{4 \; \exp H'}\right)^{-1} \tag{2.84}$$

where

$$H' = \frac{Z_0\sqrt{2(\varepsilon_r + 1)}}{119.9} + \frac{1}{2}\left(\frac{\varepsilon_r - 1}{\varepsilon_r + 1}\right)\left(\ln\frac{\pi}{2} + \frac{1}{\varepsilon_r}\ln\frac{4}{\pi}\right) \tag{2.85}$$

We may also use the following, with a slight but significant shift of changeover value
to $w/h < 1.3$ (i.e., when $Z_0 > 63 - 2\varepsilon_r$ ohms),

$$\varepsilon_{\text{eff}} = \frac{\varepsilon_r + 1}{2}\left[1 - \frac{1}{2H'}\left(\frac{\varepsilon_r - 1}{\varepsilon_r + 1}\right)\left(\ln\frac{\pi}{2} + \frac{1}{\varepsilon_r}\ln\frac{4}{\pi}\right)\right]^{-2} \tag{2.86}$$

where H' is given by Eq. (2.85) (as a function of Z_0) or, alternatively, as a function of
w/h, from Eq. (2.84):

$$H' = \ln\left[4\frac{h}{w} + \sqrt{16\left(\frac{h}{w}\right)^2 + 2}\right] \tag{2.87}$$

For wide strips (i.e., when $Z_0 < 44 - 2\varepsilon_r$ ohms),

$$\frac{w}{h} = \frac{2}{\pi}[(d_\varepsilon - 1) - \ln(2d_\varepsilon - 1)] + \frac{\varepsilon_r - 1}{\pi\varepsilon_r}\left[\ln(d_\varepsilon - 1) + 0.293 - \frac{0.517}{\varepsilon_r}\right] \tag{2.88}$$

where

$$d_\varepsilon = \frac{59.95\pi^2}{Z_0\sqrt{\varepsilon_r}}$$ (2.89)

$$\varepsilon_{\text{eff}} = \frac{\varepsilon_r + 1}{2} + \frac{\varepsilon_r - 1}{2}\left(1 + 10\frac{h}{w}\right)^{-0.555}$$ (2.90)

Alternatively, where Z_0 is known at first,

$$\varepsilon_{\text{eff}} = \frac{\varepsilon_r}{0.96 + \varepsilon_r(0.109 - 0.004\varepsilon_r)[\log(10 + Z_0) - 1]}$$ (2.91)

For microstrip lines on alumina ($\varepsilon_r = 10$), this expression appears to be accurate to $\pm 0.2\%$ over the impedance range $8 \leq Z_0 \leq 45\ \Omega$.

2.10.3   Graphical Method

For convenience, a graphical method can be used to estimate ε_{eff} and Z_0. Figures 2.24 and 2.25 show two charts that can be used to find Z_0 and ε_{eff} if w/h is known and vice-versa. Other graphs or look-up tables from static or full-wave analyses have been generated and can be found in the literature.

Example 2.4   A 50-Ω microstrip line has a stub tuner of length l as shown in Fig. 2.26. The circuit is built on Duroid 5870 with a dielectric constant of 2.3 and a thickness of 0.062 in. At 3 GHz, l is designed to be $\frac{1}{4}\lambda_g$. (a) Calculate w and l. (b) What is the VSWR? (c) At 6 GHz, what is the VSWR? Neglect the open-end discontinuity.

Solution   The stub is in shunt with the transmission line. An equivalent circuit can be given in Fig. 2.26.
(a)   From Figure 2.24,

$$\frac{w}{h} = 3 \quad \text{for 50-}\Omega \text{ line}$$

$$w = 3h = 3 \times 0.062 \text{ in.} = 0.186 \text{ in.}$$

From Fig. 2.25, we have

$$\frac{\lambda_g\sqrt{\varepsilon_r}}{\lambda_0} \approx 1.075$$

$$\lambda_g = \frac{1.075}{\sqrt{\varepsilon_r}}\lambda_0 = \frac{1.075}{\sqrt{2.3}}\frac{c}{f} = 7.09 \text{ cm}$$

$$l = \lambda_g/4 = 1.77 \text{ cm}$$

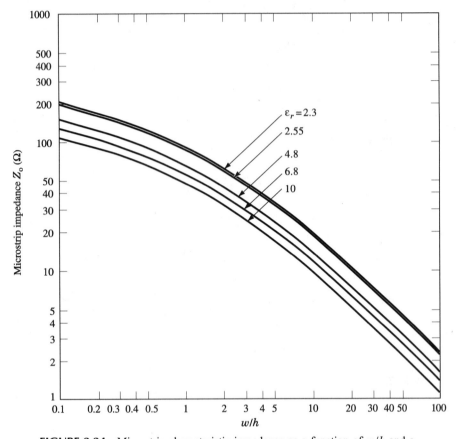

FIGURE 2.24   Microstrip characteristic impedance as a function of w/h and ε_r.

(b)

$$l = \lambda_g/4, \quad \beta l = \frac{2\pi}{\lambda_g}\frac{\lambda_g}{4} = \frac{\pi}{2}, \quad \tan \beta l = \infty$$

$$Z_{in} = Z_0 \frac{Z_L + jZ_0 \tan \beta l}{Z_0 + jZ_L \tan \beta l} \approx Z_0 \frac{Z_0}{Z_L}$$

Since $Z_L = \infty$, we have $Z_{in} = 0$. Impedance Z_{in} becomes the load to the transmission line:

$$\Gamma_L = \frac{Z_{in} - Z_0}{Z_{in} + Z_0} = -1 \qquad \text{VSWR} = \frac{1 + |\Gamma_L|}{1 - |\Gamma_L|} = \infty$$

The stub acts as a bandstop filter (or RF choke) at 3 GHz.

(c)   At 6 GHz, the same stub will become $\lambda_g/2$. Impedance $Z_{in} = \infty$. Since Z_{in} is in parallel with Z_0, the load to the transmission is Z_0. Therefore, VSWR $= 1$. In this case, the stub has no effect on the main line.   ∎

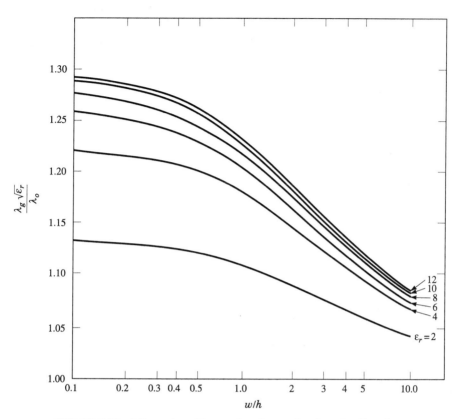

FIGURE 2.25 Microstrip guide wavelength as a function of w/h and ε_r.

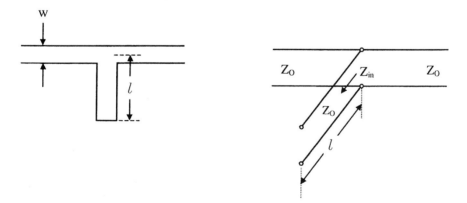

FIGURE 2.26 Microstrip line and its equivalent circuit.

2.11   WAVEGUIDES

A waveguide is a hollow metal pipe for guided wave propagation. There are two common types of waveguides: rectangular and circular [1]. The rectangular waveguide has a rectangular cross section, as shown in Fig. 2.27, and is the most widely used one. The rectangular waveguide has been used from 320 MHz up to 325 GHz with many different bands. Each band corresponds to a waveguide with certain dimensions (see Table 2.4). The WR-2300 waveguide for use at 325 MHz has internal dimensions of 23 × 11.5 in., and the WR-3 waveguide for use at 325 GHz has internal dimensions of 0.034 × 0.017 in. The commonly used X-band (WR-90) covers an 8.2–12.4-GHz frequency range with internal dimensions of 0.9 × 0.4 in. Due to their large size, high cost, heavy weight, and difficulty in integration at lower microwave frequencies, rectangular waveguides have been replaced by microstrip or coaxial lines except for very high power or high-frequency applications. In this section, the operating principle of a rectangular waveguide is briefly described.

Hollow-pipe waveguides do not support a TEM wave [1]. The types of waves that can be supported or propagated in a hollow, empty waveguide are the TE and TM modes. They are the TE modes with $E_z = 0$ and $H_z \neq 0$ and the TM modes with $H_z = 0$ and $E_z \neq 0$. Note that the TEM mode ($E_z = H_z = 0$) cannot propagate since there is no center conductor to support the current.

Figure 2.27 shows a rectangular waveguide. The analysis begins with the Helmholtz equations in the source-free region [1, 2]:

$$\nabla^2 \vec{E} + \omega^2 \mu \varepsilon \vec{E} = 0 \tag{2.92}$$

$$\nabla^2 \vec{H} + \omega^2 \mu \varepsilon \vec{H} = 0 \tag{2.93}$$

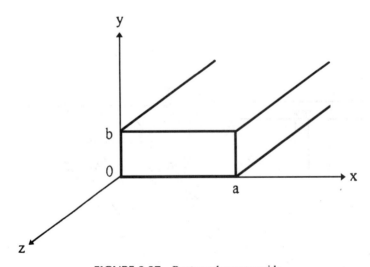

FIGURE 2.27   Rectangular waveguide.

TABLE 2.4 Rectangular Waveguide Properties

EIA WG Designation WR(·)	Recommended Operating Range for TE$_{10}$ Mode (GHz)	Cutoff Frequency for TE$_{10}$ Mode (GHz)	Theoretical CW Power Rating, Lowest to Highest Frequency (MW)	Theoretical Attenuation, Lowest to Highest Frequency (dB/100 ft)	Material Alloy	Inside Dimensions (in.)	JAN WG Designation RG(·)/U
2300	0.32–0.49	0.256	153.0–212.0	0.051–0.031	Al	23.000–11.500	
2100	0.35–0.53	0.281	120.0–173.0	0.054–0.034	Al	21.000–10.500	
1800	0.41–0.625	0.328	93.4–131.9	0.056–0.038	Al	18.000–9.000	201
1500	0.49–0.75	0.393	67.6–93.3	0.069–0.050	Al	15.000–7.500	202
1150	0.64–0.96	0.513	35.0–53.8	0.128–0.075	Al	11.500–5.750	203
975	0.75–1.12	0.605	27.0–38.5	0.137–0.095	Al	9.750–4.875	204
770	0.96–1.45	0.766	17.2–24.1	0.201–0.136	Al	7.700–3.850	205
650	1.12–1.70	0.908	11.9–17.2	0.317–0.212	Brass	6.500–3.250	69
				0.269–0.178	Al		103
510	1.45–2.20	1.157	7.5–10.7			5.100–2.550	
430	1.70–2.60	1.372	5.2–7.5	0.588–0.385	Brass	4.300–2.150	104
				0.501–0.330	Al		105
340	2.20–3.30	1.736	3.1–4.5	0.877–0.572	Brass	3.400–1.700	112
				0.751–0.492	Al		113
284	2.60–3.95	2.078	2.2–3.2	1.102–0.752	Brass	2.840–1.340	48
229	3.30–4.90	2.577	1.6–2.2	0.940–0.641	Al	2.290–1.145	75

(*continued*)

51

TABLE 2.4 (*Continued*)

EIA WG Designation WR(·)	Recommended Operating Range for TE$_{10}$ Mode (GHz)	Cutoff Frequency for TE$_{10}$ Mode (GHz)	Theoretical CW Power Rating, Lowest to Highest Frequency (MW)	Theoretical Attenuation, Lowest to Highest Frequency (dB/100 ft)	Material Alloy	Inside Dimensions (in.)	JAN WG Designation RG(·)/U
187	3.95–5.85	3.152	1.4–2.0	2.08–1.44 1.77–1.12	Brass Al	1.872–0.872	49 95
159	4.90–7.05	3.711	0.79–1.0	2.87–2.30	Brass	1.590–0.795	50
137	5.85–8.20	4.301	0.56–0.71	2.45–1.94	Al	1.372–0.622	106
112	7.05–10.00	5.259	0.35–0.46	4.12–3.21 3.50–2.74	Brass Al	1.122–0.497	51 68
90	8.20–12.40	6.557	0.20–0.29	6.45–4.48 5.49–3.383	Brass Al	0.900–0.400	52 67
75	10.00–15.00	7.868	0.17–0.23		Brass	0.750–0.375	
62	12.4–18.00	9.486	0.12–0.16	9.51–8.31 6.14–5.36	Brass Al Ag	0.622–0.311	91 107
51	15.00–22.00	11.574	0.080–0.107		Brass	0.510–0.255	
42	18.00–26.50	14.047	0.043–0.058	20.7–14.8 17.6–12.6	Brass Al	0.420–0.170	53 121
34	22.00–33.00	17.328	0.034–0.048	13.3–9.5	Ag	0.340–0.170	66

EIA	Frequency Range (GHz)	Cutoff (GHz)	Attenuation (dB/ft)	Power	Material	Attenuation (dB/ft)	Ref
28	26.50–40.00	21.081	0.022–0.031	—	Brass, Al	0.280–0.140	
22	33.00–50.00	26.342	0.014–0.020	21.9–15.0	Ag, Brass	0.224–0.112	96
19	40.00–60.00	31.357	0.011–0.015	31.0–20.9	Ag	0.188–0.094	97
15	50.00–75.00	39.863	0.0063–0.0090	—		0.148–0.074	
12	60.00–90.00	48.350	0.0042–0.0060	52.9–39.1	Brass, Ag, Brass	0.122–0.061	98
10	75.00–110.00	59.010	0.0030–0.0041	93.3–52.2	Ag	0.100–0.050	99
8	90.00–140.00	73.840	0.0018–0.0026	152–99	Ag	0.080–0.040	138
7	110.00–170.00	90.840	0.0012–0.0017	163–137	Ag	0.065–0.0325	136
5	140.00–220.00	115.750	0.00071–0.00107	308–193	Ag	0.051–0.0255	135
4	170.00–260.00	137.520	0.00052–0.00075	384–254	Ag	0.043–0.0215	137
3	220.00–325.00	173.280	0.00035–0.00047	512–348	Ag	0.034–0.0170	139

Source: M. I. Skolnik, *Radar Handbook* (New York: McGraw-Hill, 1970).

The solutions are many TE and TM modes meeting the boundary conditions.

A rectangular waveguide is normally operating in the TE_{10} mode, which is called the dominant mode. The field components for the TE_{10} mode are [1, 2]

$$H_z = A_{10} \cos\frac{\pi x}{a} e^{-j\beta_{10}z} \tag{2.94a}$$

$$H_x = \frac{j\beta_{10}a}{\pi} A_{10} \sin\frac{\pi x}{a} e^{-j\beta_{10}z} \tag{2.94b}$$

$$H_y = 0 \tag{2.94c}$$

$$E_x = 0 \tag{2.94d}$$

$$E_y = -\frac{j\omega\mu a}{\pi} A_{10} \sin\frac{\pi x}{a} e^{-j\beta_{10}x} \tag{2.94e}$$

$$E_z = 0 \tag{2.94f}$$

$$\beta_{10} = \sqrt{\omega^2\mu\varepsilon - \left(\frac{\pi}{a}\right)^2} \tag{2.94g}$$

$$\lambda_{g,10} = \frac{2\pi}{\sqrt{\omega^2\mu\varepsilon - \left(\frac{\pi}{a}\right)^2}} \tag{2.94h}$$

$$f_{c,10} = \frac{1}{2a\sqrt{\mu\varepsilon}} \tag{2.94i}$$

$$\lambda_{c,10} = 2a \tag{2.94j}$$

For an X-band (8.2–12.4 GHz) waveguide with $a = 0.9$ in. and $b = 0.4$ in., the cutoff frequency $f_{c,10} = 6.56$ GHz. The first higher order mode is the TE_{20} mode with a cutoff frequency $f_{c,20} = 13.1$ GHz. Operating between 8.2 and 12.4 GHz ensures that only one mode (i.e., the TE_{10} mode) exists.

The cutoff frequencies for higher order modes are

$$f_{c,nm} = \frac{1}{2\pi\sqrt{\mu\varepsilon}}\sqrt{\left(\frac{n\pi}{a}\right)^2 + \left(\frac{m\pi}{b}\right)^2} = \frac{c}{2\pi\sqrt{\mu_r\varepsilon_r}}\sqrt{\left(\frac{n\pi}{a}\right)^2 + \left(\frac{m\pi}{b}\right)^2} \tag{2.95}$$

where $c = 1/\sqrt{\mu_0\varepsilon_0}$ is the speed of light in a vacuum and $\mu = \mu_r\mu_0$, $\varepsilon = \varepsilon_r\varepsilon_0$. Note that $n = 1, 2, 3, \ldots$ and $m = 1, 2, 3, \ldots$ for the TM_{nm} modes and $n = 0, 1, 2, \ldots$ and $m = 0, 1, 2, \ldots$ but $n \neq m = 0$ for TE_{nm} modes. If the frequency is greater than $f_{c,nm}$, the (nm)th mode will propagate in the waveguide.

2.12 LUMPED ELEMENTS

At RF and low microwave frequencies, lumped R, L, and C elements can be effectively used in the circuit design if the length of the element is very small in comparison to the wavelength. At low frequencies, wavelengths are big and, consequently, stubs are big. Lumped elements have the advantages of smaller size and wider bandwidth as compared to distributive elements (stubs, line sections, etc.). At high frequencies, however, distributive circuits are used because of their lower loss (or higher Q), and the size advantage of lumped elements is no longer a significant factor.

Although some simple closed-form design equations are available for the lumped-element design, an accurate design would require the use of computer-aided design (CAD). Lumped-element components normally exhibit undesirable effects such as spurious resonances, fringing fields, parasitic capacitance and inductance, parasitic resistance and loss, and various perturbations. Figure 2.28 shows different types of lumped R, C, and L elements [10]. With the advent of new photolithographic techniques, the fabrication of lumped elements can now be extended to 60 GHz and beyond.

2.13 IMPEDANCE MATCHING NETWORKS

In Section 2.4, when a load is connected to a transmission line, strong reflection can result if the load impedance differs from the line's characteristic impedance. To reduce the reflection, a lossless impedance matching network is inserted between the load and the line, as shown in Fig. 2.29. The matching network will transform the load impedance into an input impedance that matches the line's characteristic impedance. This type of matching can be classified as one-port impedance matching.

For a two-port component or device such as a filter, amplifier, as multiplier, both input matching and output matching networks will be required (Fig. 2.30). For example, in a transistor amplifier design for maximum gain, one needs to match Z_0 to Z_s at the input port and Z_0 to Z_L at the output port, respectively. Impedances Z_s and Z_L are obtained from Γ_s and Γ_L, which are designed equal to S_{11}^* and S_{22}^* for maximum power transfer. Here the asterisks denote the complex conjugate quantities, S_{11} and S_{22} being the S-parameters of the transistor device. Γ_s and Γ_L are the reflection coefficients looking into the source and the load, respectively, from the transistor.

Several methods can be used for impedance matching:

1. Matching stubs (shunt or series, single or multiple)
2. Quarter-wavelength transformers (single or multiple)
3. Lumped elements
4. Combinations of the above

The multiple-section transformers and the tapered transmission lines are generally used for broadband impedance matching. Computer-aided design using commercially available software can be utilized to facilitate design.

2.13.1 Matching Stubs

One can use a single or double stub to accomplish impedance matching. Figure 2.31 shows a single-shunt stub matching network that is used to match Z_L to Z_0. For a

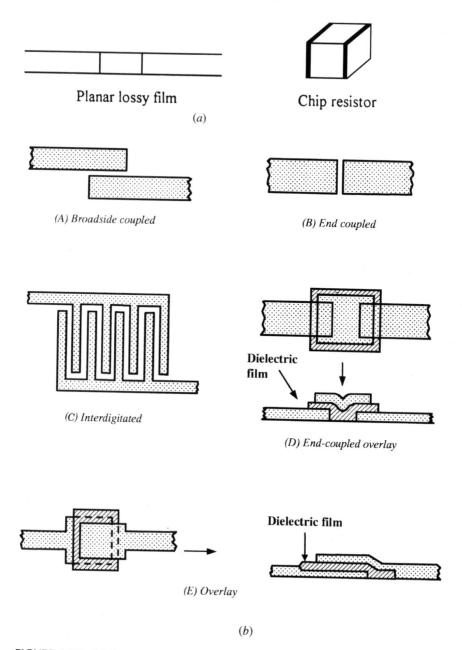

Planar lossy film

(a)

Chip resistor

(A) Broadside coupled

(B) End coupled

(C) Interdigitated

Dielectric film

(D) End-coupled overlay

(E) Overlay

Dielectric film

(b)

FIGURE 2.28   (a) Some planar resistor configurations. (b) Some planar capacitor configurations [10]. (Permission from H. W. Sams)

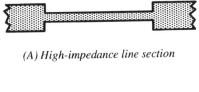

(A) High-impedance line section

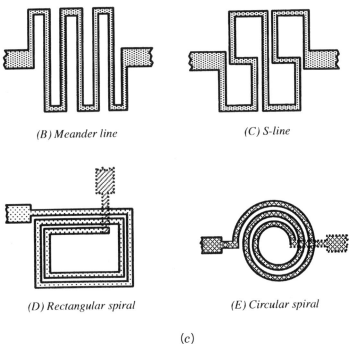

(B) Meander line   *(C) S-line*

(D) Rectangular spiral   *(E) Circular spiral*

(c)

FIGURE 2.28   *(c)* Some planar inductor configurations [10]. (Permission from H. W. Sams)

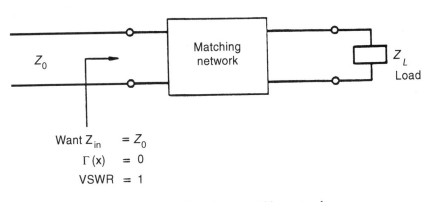

FIGURE 2.29   Impedance matching network.

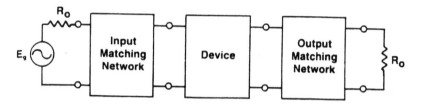

FIGURE 2.30   Two-port matching network ($R_0 = Z_0$).

shunt stub, a Y-chart is used for the design. A Z-chart is more convenient for the series stub design. One needs to design l_1 and l_2 such that $\bar{Y}_{in} = 1$. Although a shorted stub is employed in Fig. 2.31, an open stub could also be used.

To achieve the matching, l_1 is designed such that $\bar{Y}_A = 1 + jb$, and l_2 is designed such that $\bar{Y}_t = -jb$. Thus,

$$\bar{Y}_{in} = \bar{Y}_t + \bar{Y}_A = 1 \tag{2.96a}$$

$$\bar{Y}_A = \frac{\bar{Y}_L + j \tan \beta l_1}{1 + j\bar{Y}_L \tan \beta l_1} = 1 + jb \tag{2.96b}$$

$$\bar{Y}_t = -j \cot \beta l_2 = -jb \tag{2.96c}$$

The design can be easily accomplished by using a Smith chart [2].

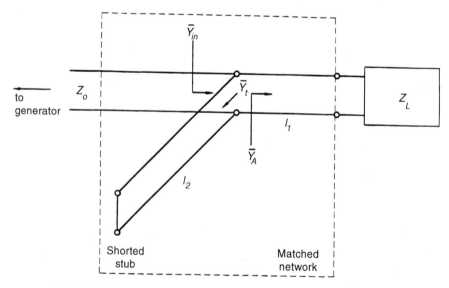

FIGURE 2.31   Matching network using a single stub to match Z_L to Z_0.

2.13.2 Single-Section Quarter-Wavelength Transformer

A quarter-wavelength transformer is a convenient way to accomplish impedance matching. For the resistive load shown in Fig. 2.32, only a quarter-wavelength transformer alone can achieve the matching. This can be seen from the following. To achieve impedance matching, we want $Z_{in} = Z_0$. For a line of length l_T and characteristic impedance of Z_{0T}, the input impedance is

$$Z_{in} = Z_{0T} \frac{Z_L + jZ_{0T} \tan \beta l_T}{Z_{0T} + jZ_L \tan \beta l_T} \tag{2.97}$$

Since $l_T = \frac{1}{4}\lambda_g$, $\beta l_T = \frac{1}{2}\pi$, and $\tan \beta l_T = \infty$, we have

$$Z_{in} = \frac{Z_{0T}^2}{Z_L} = \frac{Z_{0T}^2}{R_L} = Z_0 \tag{2.98}$$

Thus

$$Z_{0T} = \sqrt{Z_0 R_L} \tag{2.99}$$

Equation (2.99) implies that the characteristic impedance of the quarter-wavelength section should be equal to $\sqrt{Z_0 R_L}$ if a quarter-wavelength transformer is used to match R_L to Z_0. Since the electrical length is a function of frequency, the matching is good only for a narrow frequency range. The same argument applies to stub matching networks. To accomplish wideband matching, a multiple-section matching network is required.

For a complex load $Z_L = R_L + jX_L$, one needs to use a single stub tuner to tune out the reactive part, then use a quarter-wavelength transformer to match the remaining real part of the load. Figure 2.33 shows this matching arrangement.

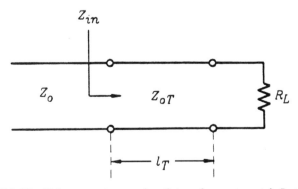

FIGURE 2.32 Using a quarter-wavelength transformer to match R_L to Z_0.

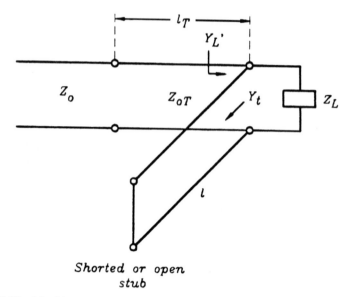

FIGURE 2.33 Matching network using a quarter-wavelength transformer together with a single-stub tuner.

Alternatively, one can transform Z_L to a real value through a line l followed by a quarter-wavelength transformer.

2.13.3 Lumped Elements

At RF and low microwave frequencies, lumped elements (L, C) can be effectively used to accomplish impedance matching. Lumped elements have the advantages of small size and wider bandwidth as compared to distributive circuits. At high frequencies, distributive circuits such as tuning stubs become small and have the advantage of lower loss.

Many combinations of LC circuits can be used to match Z_L to Z_0, as shown in Fig. 2.34. Adding a series reactance to a load produces a motion along a constant-resistance circle in the Z Smith chart. As shown in Fig. 2.35, if the series reactance is an inductance, the motion is upward since the total reactance is increased. If it is a capacitance, the motion is downward. Adding a shunt reactance to a load produces a motion along a constant-conductance circle in the rotated Y-chart. As shown in Fig. 2.36, if the shunt element is an inductor, the motion is upward. If the shunt element is a capacitor, the motion is downward. Using the combination of constant-resistance circles and constant-conductance circles, one can move $\bar{Z}_L$ (or $\bar{Y}_L$) to the center of the Z–Y chart.

Consider a case using a series L and a shunt C to match Z_L to Z_0, as shown in Fig. 2.37. The procedure is as follows:

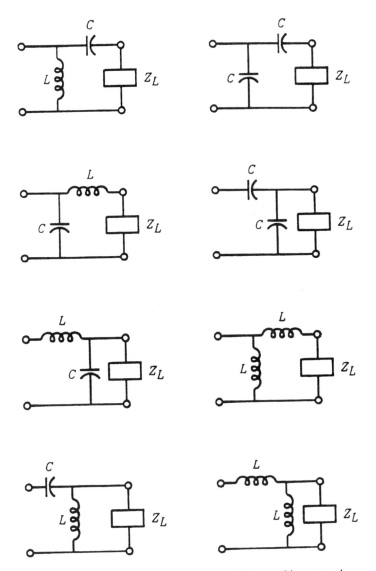

FIGURE 2.34 *LC* lumped elements used for matching network.

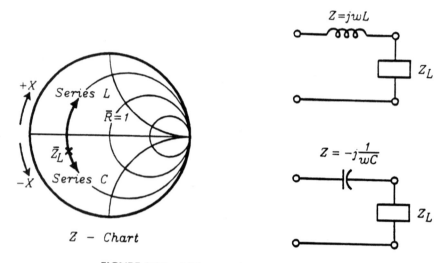

FIGURE 2.35   Adding a series reactance to a load.

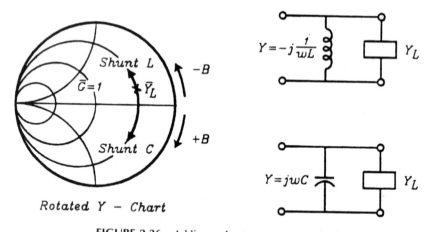

FIGURE 2.36   Adding a shunt reactance to a load.

1. Locate A from $\bar{Z}_L$ in the Z-chart.
2. Move A to B along the constant-resistance circle. Point B is located on the constant-conductance circle passing through the center (i.e., $\bar{G} = 1$ circle).
3. Move from B to C along the $\bar{G} = 1$ circle. Point C is the center of the chart.
4. From the Z-chart and rotated Y-chart, one can find values of L and C, respectively.

Examples can be found in other circuit design books [1, 2].

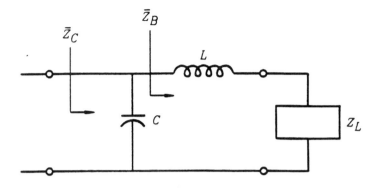

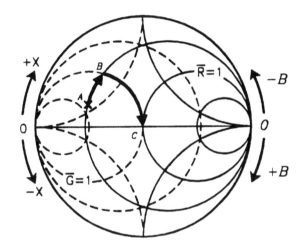

FIGURE 2.37   Using a series L and a shunt C to match Z_L to Z_0.

PROBLEMS

2.1   Convert the following quantities into dB or dBm: (a) 1500, (b) $\frac{1}{2}$, (c) 0.00004321, (d) 0.01 mW, and (e) 20 mW.

2.2   What is the output power of the system shown in Fig. P2.2 ($L = $ loss, $G = $ gain)?

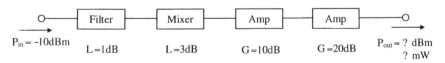

FIGURE P2.2

2.3   A load is connected to a transmission line. Calculate the percentage of the incident power being reflected if the VSWR of the load is: (a) 1.1, (b) 1.2, (c) 2.0, (d) 3.0, and (e) 10.0.

2.4   An air-filled transmission line with a length l of 20 cm is connected to a load of 100 Ω. Calculate the input impedance for the following frequencies: (a) 10 kHz, (b) 1 MHz, (c) 100 MHz, (d) 3 GHz, and (e) 10 GHz. (Note that $c = f \lambda_0$ and $\lambda_g = \lambda_0$, $Z_0 = 50 \, \Omega$.)

2.5   A transmission line with $Z_0 = 50 \, \Omega$ is connected to a load of 100 Ω. (a) Calculate the reflection coefficient Γ_L and VSWR; (b) find the percentage of incident power being reflected and the return loss; (c) plot $|V(x)|$ as a function of x; and (d) find the input impedance at $x = -0.5\lambda_g$ and $x = -0.25\lambda_g$.

2.6   An antenna with a VSWR of 1.5 is connected to a 50-Ω transmission line. What is the reflected power in watts if the input power is 10 W?

2.7   An antenna is connected to a 100-W transmitter through a 50-Ω coaxial cable (i.e., characteristic impedance of 50 Ω; see Fig. P2.7). The input impedance of the antenna is 45 Ω. The transmitter output impedance is perfectly matched to the cable. Calculate (a) the VSWR at the antenna input port, (b) the return loss at the input port, and (c) the reflected power in watts at the input port.

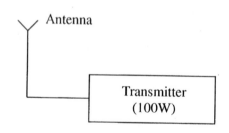

FIGURE P2.7

2.8   A coaxial line is filled with a material ($\varepsilon_r = 2$, $\mu_r = 1$) has an inner conductor diameter of 2 mm and an outer conductor diameter of 10 mm. Calculate (a) the characteristic impedance and (b) the guided wavelength at 10 GHz.

2.9   Repeat Problem 2.8 for a microstrip line with $w = 0.075$ in., $h = 0.025$ in., and $\varepsilon_r = 2.3$, $\mu_r = 1$.

2.10 A microstrip line is built on a substrate with a dielectric constant of 10. The substrate thickness is 0.025 in. Calculate the line width and guide wavelength for a 50-Ω line operating at 10 GHz.

2.11 A microstrip line is built on a substrate with a dielectric constant of 10. The substrate thickness is 0.010 in. Calculate the line width and guide wavelength for a 50-Ω line operating at 10 GHz. What is the effective dielectric constant for this case?

2.12 A coaxial line is filled with a material of $\varepsilon_r = 4$ and $\mu_r = 1$. The coaxial line has an inner conductor radius of 2 mm and an outer conductor radius of 10.6 mm. The coaxial is connected to a 25-Ω load. Calculate (a) the characteristic impedance of the coaxial line, (b) the percentage of incident power being delivered to the load, and (c) the input impedance at 6.25 cm away from the load at 3 GHz.

2.13 A microstrip line is built on a substrate with a thickness of 0.030 in. and a relative dielectric constant of 10. The line width is 0.120 in. The operating frequency is 1 GHz. The line is connected to a load of 30 Ω. Calculate (a) the guide wavelength, (b) the VSWR, (c) the percentage of incident power being reflected, and (d) the return loss in decibels.

2.14 A microstrip line is built on a substrate with a thickness of 0.030 in. with a line width of 0.090 in. The dielectric constant of the substrate is 2.3 and the operating frequency is 3 GHz. The line is connected to a antenna with an input impedance of 100 Ω. Calculate (a) the VSWR, (b) the percentage of incident power being reflected, and (c) the return loss in decibels. (d) If a quarter-wavelength transformer is used to match the line to the antenna, what is the characteristic impedance of the transformer?

2.15 A 50-Ω microstrip line is built on a Duroid 6010 substrate ($\varepsilon_r = 10$) with a thickness of 0.030 in. The line is connected to a 10-Ω load as shown in Fig. P2.15. Determine (a) the reflection coefficient Γ_L, VSWR, percentage of incident power being reflected, and return loss. (b) If a quarter-wavelength transformer is used to match this load as shown in the figure, design w_1, w_2, and l_1 in millimeters. The operating frequency is 3 GHz.

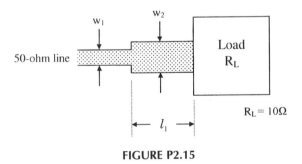

FIGURE P2.15

REFERENCES

1. R. E. Collin, *Foundations for Microwave Engineering*, McGraw-Hill, New York, 1st ed. 1966, 2nd ed. 1992.

2. K. Chang, *Microwave Solid-State Circuits and Applications*, John Wiley & Sons, New York, 1994.

3. E. A. Wolff and R. Kaul, *Microwave Engineering and Systems Applications*, John Wiley & Sons, New York, 1988.

4. H. A. Wheeler, "Transmission-Line Properties of Parallel Strips Separated by a Dielectric Sheet," *IEEE Trans. Microwave Theory Tech.*, Vol. MTT-13, pp. 172–185, 1965.

5. H. E. Brenner, "Numerical Solution of TEM-Line Problems Involving Inhomogeneous Media," *IEEE Trans. Microwave Theory Tech.*, Vol. MTT-15, pp. 485–487, 1967.

6. H. E. Stinehelfer, "An Accurate Calculation of Uniform Microstrip Transmission-Lines," *IEEE Trans. Electron Devices*, Vol. ED-15, pp. 501–506, 1968.

7. E. Yamashita and R. Mittra, "Variational Method for the Analysis of Microstrip Lines," *IEEE Trans. Microwave Theory Tech.*, Vol. MTT-16, pp. 529–535, 1968.

8. B. Bhat and S. K. Koul, "Unified Approach to Solve a Class of Strip and Microstrip-Like Transmission-Lines," *IEEE Trans. Microwave Theory Tech.*, Vol. MTT-30, pp. 679–686, 1982.

9. T. C. Edwards, *Foundations for Microstrip Circuit Design*, John Wiley & Sons, New York, 1st ed., 1981, 2nd ed., 1991, Chs. 3–5.

10. R. A. Pucel, "Technology and Design Considerations of Monolithic Microwave Integrated Circuits," in D. K. Ferry, Ed., *Gallium Arsenide Technology*, H. W. Sams Co., Indianapolis, IN, 1985, pp. 189–248.

CHAPTER THREE

Antenna Systems

3.1 INTRODUCTION

The study of antennas is very extensive and would need several texts to cover adequately. In this chapter, however, a brief description of relevant performances and design parameters will be given for introductory purposes.

An antenna is a component that radiates and receives the RF or microwave power. It is a reciprocal device, and the same antenna can serve as a receiving or transmitting device. Antennas are structures that provide transitions between guided and free-space waves. Guided waves are confined to the boundaries of a transmission line to transport signals from one point to another [1], while free-space waves radiate unbounded. A transmission line is designed to have very little radiation loss, while the antenna is designed to have maximum radiation. The radiation occurs due to discontinuities (which cause the perturbation of fields or currents), unbalanced currents, and so on.

The antenna is a key component in any wireless system, as shown in Fig. 3.1. The RF/microwave signal is transmitted to free space through the antenna. The signal propagates in space, and a small portion is picked up by a receiving antenna. The signal will then be amplified, downconverted, and processed to recover the information.

There are many types of antennas; Fig. 3.2 gives some examples. They can be classified in different ways. Some examples are:

1. Shapes or geometries:
    a. Wire antennas: dipole, loop, helix
    b. Aperture antennas: horn, slot
    c. Printed antennas: patch, printed dipole, spiral

2. Gain:
   a. High gain: dish
   b. Medium gain: horn
   c. Low gain: dipole, loop, slot, patch
3. Beam shapes:
   a. Omnidirectional: dipole
   b. Pencil beam: dish
   c. Fan beam: array

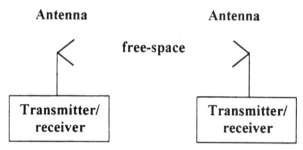

FIGURE 3.1 Typical wireless system.

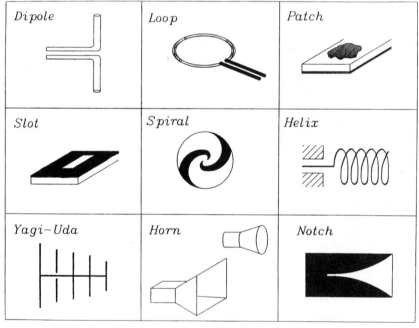

FIGURE 3.2 Various antennas [2].

4. Bandwidth:
   a. Wide band: log, spiral, helix
   b. Narrow band: patch, slot

Since antennas interface circuits to free space, they share both circuit and radiation qualities. From a circuit point of view, an antenna is merely a one-port device with an associated impedance over frequency. This chapter will describe some key antenna properties, followed by the designs of various antennas commonly used in wireless applications.

3.2 ISOTROPIC RADIATOR AND PLANE WAVES

An isotropic radiator is a theoretical point antenna that cannot be realized in practice. It radiates energy equally well in all directions, as shown in Fig. 3.3. The radiated energy will have a spherical wavefront with its power spread uniformly over the surface of a sphere. If the source transmitting power is P_t, the power density P_d in watts per square meters at a distance R from the source can be calculated by

$$P_d = \frac{P_t}{4\pi R^2} \tag{3.1}$$

Although the isotropic antenna is not practical, it is commonly used as a reference with which to compare other antennas.

At a distance far from the point source or any other antenna, the radiated spherical wave resembles a uniform plane wave in the vicinity of the receiving antenna. This can be understood from Fig. 3.4. For a large R, the wave can be approximated by a uniform plane wave. The electric and magnetic fields for plane waves in free space can be found by solving the Helmholtz equation

$$\nabla^2 \vec{E} + k_0^2 \vec{E} = 0 \tag{3.2}$$

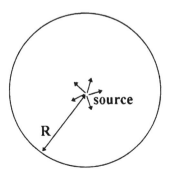

FIGURE 3.3 Isotropic radiator.

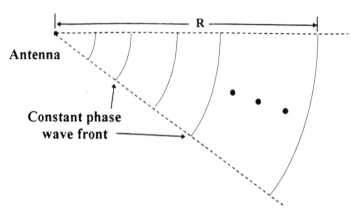

FIGURE 3.4   Radiation from an antenna.

where $k_0 = 2\pi/\lambda_0$. The solution is [1]

$$\vec{E} = \vec{E}_0 e^{-j\vec{k}_0 \cdot \vec{r}}$$ (3.3)

The magnetic field can be found from the electric field using the Maxwell equation, given by

$$\vec{H} = -\frac{1}{j\omega\mu_0}\nabla \times \vec{E} = \sqrt{\frac{\varepsilon_0}{\mu_0}}\,\hat{n} \times \vec{E}$$ (3.4)

where μ_0 is the free-space permeability, ε_0 is the free-space permittivity, ω is the angular frequency, and k_0 is the propagation constant. Here, $\vec{k}_0 = \hat{n}k_0$, and $\hat{n}$ is the unit vector in the wave propagation direction, as shown in Fig. 3.5. The vector E_0 is perpendicular to the direction of the propagation, and H is perpendicular to $\vec{E}$ and $\hat{n}$. Both $\vec{E}$ and $\vec{H}$ lie in the constant-phase plane, and the wave is a TEM wave.

The intrinsic impedance of free space is defined as

$$\eta_0 = \frac{E}{H} = \sqrt{\frac{\mu_0}{\varepsilon_0}} = 120\pi \text{ or } 377 \ \Omega$$ (3.5)

The time-averaged power density in watts per square meters is given as

$$P_d = \left|\frac{1}{2}\vec{E} \times \vec{H}^*\right| = \frac{1}{2}\frac{E^2}{\eta_0}$$ (3.6)

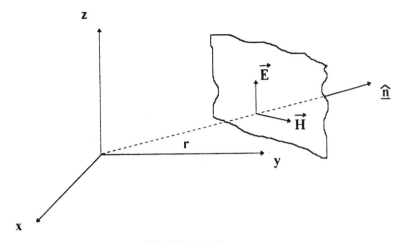

FIGURE 3.5 Plane wave.

where the asterisk denotes the complex conjugate quantity. By equating Eq. (3.1) and Eq. (3.6), one can find the electric field at a distance R from the isotropic antenna as

$$E = \frac{\sqrt{60P_t}}{R} = \sqrt{2}E_{rms} \qquad (3.7)$$

where E is the peak field magnitude and E_{rms} is the root-mean-square (rms) value.

3.3 FAR-FIELD REGION

Normally, one assumes that the antenna is operated in the far-field region, and radiation patterns are measured in the far-field region where the transmitted wave of the transmitting antenna resembles a spherical wave from a point source that only locally resembles a uniform plane wave. To derive the far-field criterion for the distance R, consider the maximum antenna dimension to be D, as shown in Fig. 3.6. We have

$$\begin{aligned} R^2 &= (R - \Delta l)^2 + (\tfrac{1}{2}D)^2 \\ &= R^2 - 2R\,\Delta l + (\Delta l)^2 + (\tfrac{1}{2}D)^2 \end{aligned} \qquad (3.8)$$

For $R \gg \Delta l$, Eq. (3.8) becomes

$$2R\,\Delta l \approx \tfrac{1}{4}D^2 \qquad (3.9)$$

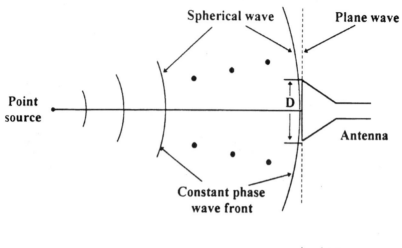

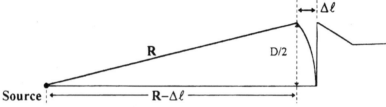

FIGURE 3.6 Configuration used for calculation of far-field region criterion.

Therefore

$$R = \frac{D^2}{8\,\Delta l} \tag{3.10}$$

If we let $\Delta l = \frac{1}{16}\lambda_0$, which is equivalent to 22.5° phase error, be the criterion for far-field operation, we have

$$R_{\text{far-field}} = \frac{2D^2}{\lambda_0} \tag{3.11}$$

where λ_0 is the free-space wavelength. The condition for far-field operation is thus given by

$$R \geq \frac{2D^2}{\lambda_0} \tag{3.12}$$

It should be noted that other criteria could also be used. For example, if $\Delta l = \frac{1}{32}\lambda_0$ or 11.25° phase error, the condition will become $R \geq 4D^2/\lambda_0$ for far-field operation.

3.4 ANTENNA ANALYSIS

To analyze the electromagnetic radiation of an antenna, one needs to work in spherical coordinates. Considering an antenna with a volume V and current $\vec{J}$ flowing in V, as shown in Fig. 3.7, the electric and magnetic fields can be found by solving the inhomogeneous Helmholtz equation [1]:

$$\nabla^2\vec{A} + k_0^2\vec{A} = -\mu\vec{J} \tag{3.13}$$

where $\vec{A}$ is the vector potential, defined as

$$\vec{B} = \nabla \times \vec{A} = \mu_0\vec{H} \tag{3.14}$$

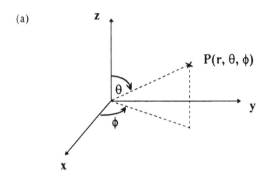

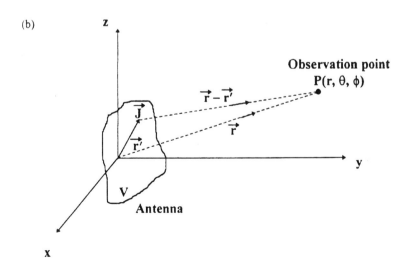

FIGURE 3.7   Antenna analysis: (*a*) spherical coordinates; (*b*) antenna and observation point.

The radiation is due to the current flow on the antenna, which contributes to a vector potential at point $P(r, \theta, \phi)$. This vector potential is the solution of Eq. (3.13), and the result is given by [1]

$$\vec{A}(\vec{r}) = \frac{\mu}{4\pi} \int_V \vec{J}(\vec{r}') \frac{e^{-jk_0|\vec{r}-\vec{r}'|}}{|\vec{r} - \vec{r}'|} dV' \qquad (3.15)$$

where r' is the source coordinate and r is the observation point coordinate. The integral is carried over the antenna volume with the current distribution multiplied by the free-space Green's function, defined by

$$\text{Free-space Green's function} = \frac{e^{-jk_0|\vec{r}-\vec{r}'|}}{|\vec{r} - \vec{r}'|} \qquad (3.16)$$

If the current distribution is known, then $\vec{A}(\vec{r})$ can be determined. From $\vec{A}(\vec{r})$, one can find $\vec{H}(\vec{r})$ from Eq. (3.14) and thus the electric field $\vec{E}(\vec{r})$. However, in many cases, the current distribution is difficult to find, and numerical methods are generally used to determine the current distribution.

3.5   ANTENNA CHARACTERISTICS AND PARAMETERS

There are many parameters used to specify and evaluate a particular antenna. These parameters provide information about the properties and characteristics of an antenna. In the following, these parameters will be defined and described.

3.5.1   Input VSWR and Input Impedance

As the one-port circuit, an antenna is described by a single scattering parameter S_{11} or the reflection coefficient Γ, which gives the reflected signal and quantifies the impedance mismatch between the source and the antenna. From Chapter 2, the input VSWR and return loss are given by

$$\text{VSWR} = \frac{1 + |\Gamma|}{1 - |\Gamma|} \qquad (3.17)$$

$$\text{RL in dB} = -20 \log|\Gamma| \qquad (3.18)$$

The optimal VSWR occurs when $|\Gamma| = 0$ or VSWR $= 1$. This means that all power is transmitted to the antenna and there is no reflection. Typically, VSWR $\leqslant 2$ is acceptable for most applications. The power reflected back from the antenna is $|\Gamma|^2$ times the power available from the source. The power coupled to the antenna is $(1 - |\Gamma|^2)$ times the power available from the source.

The input impedance is the one-port impedance looking into the antenna. It is the impedance presented by the antenna to the transmitter or receiver connected to it. The input impedance can be found from Γ by

$$Z_{\text{in}} = Z_0 \frac{1 + \Gamma}{1 - \Gamma} \tag{3.19}$$

where Z_0 is the characteristic impedance of the connecting transmission line. For a perfect matching, the input impedance should be equal to Z_0.

3.5.2 Bandwidth

The bandwidth of an antenna is broadly defined as the range of frequencies within which the performance of the antenna, with respect to some characteristic, conforms to a specified standard. In general, bandwidth is specified as the ratio of the upper frequency to the lower frequency or as a percentage of the center frequency. Since antenna characteristics are affected in different ways as frequency changes, there is no unique definition of bandwidth. The two most commonly used definitions are pattern bandwidth and impedance bandwidth.

The power entering the antenna depends on the input impedance locus of the antenna over the frequencies. Therefore, the impedance bandwidth (BW) is the range of frequencies over which the input impedance conforms to a specified standard. This standard is commonly taken to be VSWR ≤ 2 (or $|\Gamma| \leq \frac{1}{3}$) and translates to a reflection of about 11% of input power. Figure 3.8 shows the bandwidth definition [2]. Some applications may require a more stringent specification, such as a VSWR of 1.5 or less. Furthermore, the operating bandwidth of an antenna could be smaller than the impedance bandwidth, since other parameters (gain, efficiency, patterns, etc.) are also functions of frequencies and may deteriorate over the impedance bandwidth.

3.5.3 Power Radiation Patterns

The power radiated (or received) by an antenna is a function of angular position and radial distance from the antenna. At electrically large distances, the power density drops off as $1/r^2$ in any direction. The variation of power density with angular position can be plotted by the radiation pattern. At electrically large distances (i.e., far-field or plane-wave regions), the patterns are independent of distance.

The complete radiation properties of the antenna require that the electric or magnetic fields be plotted over a sphere surrounding the antenna. However, it is often enough to take principal pattern cuts. Antenna pattern cuts are shown in Fig. 3.9. As shown, the antenna has E- and H-plane patterns with co- and cross-polarization components in each. The E-plane pattern refers to the plane containing the electric field vector (E_θ) and the direction of maximum radiation. The parameter E_ϕ is the cross-polarization component. Similarly, the H-plane pattern contains the magnetic field vector and the direction of maximum radiation. Figure 3.10 shows an antenna pattern in either the E- or H-plane. The pattern contains information about half-power beamwidth, sidelobe levels, gain, and so on.

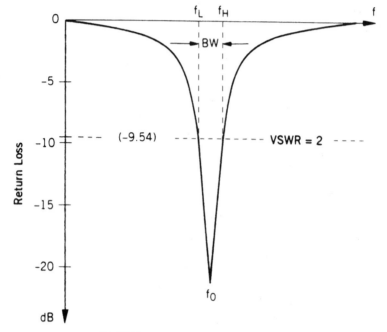

FIGURE 3.8 VSWR = 2 bandwidth [2].

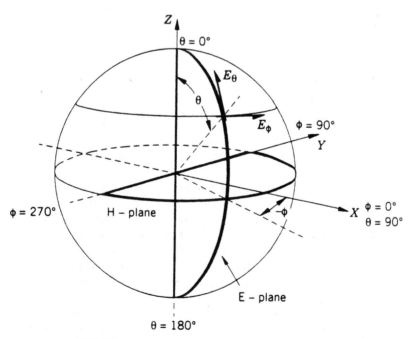

FIGURE 3.9 Antenna pattern coordinate convention [2].

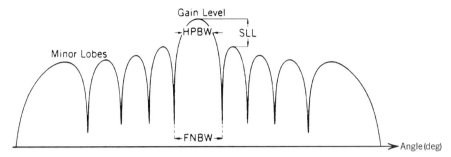

FIGURE 3.10 Antennas pattern characteristics [2].

3.5.4 Half-Power Beamwidth and Sidelobe Level

The half-power beamwidth (HPBW) is the range in degrees such that the radiation drops to one-half of (or 3 dB below) its maximum. The sidelobes are power radiation peaks in addition to the main lobe. The sidelobe levels (SLLs) are normally given as the number of decibels below the main-lobe peak. Figure 3.10 [2] shows the HPBW and SLLs. Also shown is FNBW, the first-null beamwidth.

3.5.5 Directivity, Gain, and Efficiency

The directivity D_{max} is defined as the value of the directive gain in the direction of its maximum value. The directive gain $D(\theta, \phi)$ is the ratio of the Poynting power density $S(\theta, \phi)$ over the power density radiated by an isotropic source. Therefore, one can write

$$D(\theta, \phi) = \frac{S(\theta, \phi)}{P_t/4\pi R^2} \tag{3.20}$$

$$D_{max} = \frac{\text{maximum of } S(\theta, \phi)}{P_t/4\pi R^2} \tag{3.21}$$

where $\vec{S}(\theta, \phi) = \frac{1}{2} \text{Re}[\vec{E} \times \vec{H}^*]$.

The directivity of an isotropic antenna equals to 1 by definition, and that of other antennas will be greater than 1. Thus, the directivity serves as a figure of merit relating the directional properties of an antenna relative to those of an isotropic source.

The gain of an antenna is the directivity multiplied by the illumination or aperture efficiency of the antenna to radiate the energy presented to its terminal:

$$\text{Gain} = G = \eta D_{max} \tag{3.22}$$

$$\eta = \text{efficiency} = \frac{P_{rad}}{P_{in}} = \frac{P_{rad}}{P_{rad} + P_{loss}} \tag{3.23}$$

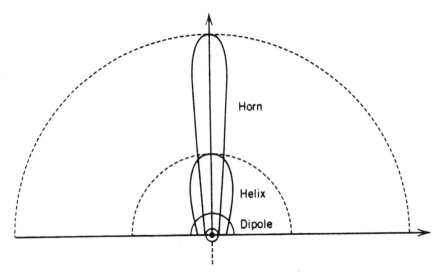

FIGURE 3.11 Gain comparison [2].

where P_{rad} is the actual power radiated, P_{in} is the power coupled into the antenna, and P_{loss} is the power lost in the antenna. The losses could include ohmic or conductor loss, dielectric loss, and so on. In general, the narrower the beamwidth, the higher the gain. Figure 3.11 gives a comparison of gain for three different antennas. From Eqs. (3.21) and (3.22), the radiated power density in the direction of its maximum value is $P_{d,\max} = G(P_t/4\pi R^2)$.

3.5.6 Polarization and Cross-Polarization Level

The polarization of an antenna is the polarization of the electric field of the radiated wave. Antennas can be classified as linearly polarized (LP) or circularly polarized (CP). The polarization of the wave is described by the tip of the E-field vector as time progresses. If the locus is a straight line, the wave is linearly polarized. If the locus is a circle, the wave is circularly polarized. Ideally, linear polarization means that the electric field is in only one direction, but this is seldom the case. For linear polarization, the cross-polarization level (CPL) determines the amount of polarization impurity. As an example, for a vertically polarized antenna, the CPL is due to the E-field existing in the horizontal direction. Normally, CPL is a measure of decibels below the copolarization level.

3.5.7 Effective Area

The effective area (A_e) is related to the antenna gain by

$$G = \frac{4\pi}{\lambda_0^2} A_e \tag{3.24}$$

It is easier to visualize the concept of effective area when one considers a receiving antenna. It is a measure of the effective absorbing area presented by an antenna to an incident wave [3]. The effective area is normally proportional to, but less than, the physical area of the antenna.

3.5.8 Beam Efficiency

Beam efficiency is another frequently used parameter to gauge the performance of an antenna. Beam efficiency is the ratio of the power received or transmitted within a cone angle to the power received or transmitted by the whole antenna. Thus, beam efficiency is a measure of the amount of power received or transmitted by minor lobes relative to the main beam.

3.5.9 Back Radiation

The back radiation is directed to the backside of an antenna. Normally it is given as the back-to-front ratio in decibels.

3.5.10 Estimation of High-Gain Antennas

There are some convenient formulas for making quick estimates of beamwidths and gains of electrically large, high gain antennas. A convenient equation for predicting a 3-dB beamwidth is [3]:

$$BW = K_1 \frac{\lambda_0}{D} \qquad (3.25)$$

where D is the aperture dimension in the plane of the pattern. For a rough estimate, one can use $K_1 = 70°$. For an example, if the length of an antenna is 10 cm, the beamwidth at 30 GHz, in the plane of length, is $7°$.

A convenient equation for predicting gain is given in reference [3]:

$$G = \frac{K_2}{\theta_1 \theta_2} \qquad (3.26)$$

where K_2 is a unitless constant and θ_1 and θ_2 are the 3-dB beamwidths in degrees in the two orthogonal principal planes. The correct K_2 value depends on antenna efficiency, but a rough estimate can be made with $K_2 = 30,000$.

Example 3.1 The E-plane pattern of an eight-element microstrip patch antenna array is shown in Fig. 3.12 [4]. Describe the characteristics of this pattern.

Solution From the pattern shown in Fig. 3.12, it can be seen that the gain is 11.4 dB. The half-power beamwidth is about $22.2°$. The cross-polarization radiation level is over 26 dB below the copolarization radiation in the main beam. The first

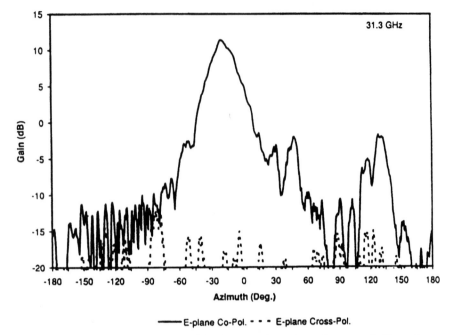

FIGURE 3.12   *E*-plane pattern of an eight-element microstrip patch antenna fed by an image line operating at 31.3 GHz. (From reference [4], with permission from IEEE.)

SLL is about 14 dB below the main lobe. The maximum back radiation occurs at around 135° with a level of about 13 dB below the main-lobe peak.   ∎

3.6   MONOPOLE AND DIPOLE ANTENNAS

The monopole and dipole antennas are commonly used for broadcasting, cellular phones, and wireless communications due to their omnidirective property. Figure 3.13 shows some examples. A monopole together with its image through a metal or ground plane has radiation characteristics similar to a dipole. A dipole with a length l is approximately equivalent to a monopole with a length of $\frac{1}{2}l$ placed on a metal or ground plane. For a dipole with a length $l < \lambda_0$, the *E*-plane radiation pattern is a doughnut shape with a hole or figure-eight shape, as shown in Fig. 3.14. The maximum radiation occurs when $\theta = 90°$ and there is no radiation at $\theta = 0°$. The *H*-plane radiation pattern is a circle, which means the antenna radiates equally in all ϕ directions. Therefore, it is nondirective in the ϕ direction. Since it is only directive in the θ direction, it is called an omnidirective antenna. Because it is nondirective in the ϕ direction, the antenna can receive a signal coming from any direction in the *H*-plane. This makes the antenna useful for broadcast and wireless applications. The antenna has no gain in the *H*-plane, that is, $G = 1$. In the *E*-plane pattern, the gain is fairly low since it has a broad beamwidth.

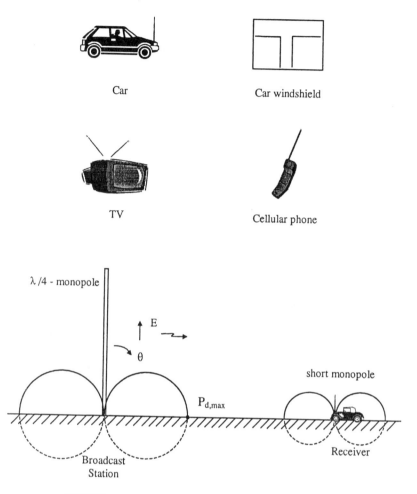

FIGURE 3.13 Examples of dipole and monopole antennas.

By assuming a sinusoidal current distribution on the dipole, the radiated fields can be found from Eqs. (3.14) and (3.15) [5]. Figure 3.15 shows the pattern plots for different dipole length l's [3]. It can be seen that the pattern deteriorates from its figure-eight shape when $l > \lambda_0$. In most cases, a dipole with $l < \lambda_0$ is used.

When $l = \frac{1}{2}\lambda_0$, it is called a half-wave dipole. A half-wave dipole (or a quarter-wavelength monopole) has an input impedance approximately equal to 73 Ω and an antenna gain of 1.64. For a very short dipole with $l \ll \lambda_0$, the input impedance is very small and difficult to match. It has low efficiency, and most power is wasted. The gain for a short dipole is approximately equal to 1.5 (or 1.7 dB).

In practice, the antenna is always fed by a transmission line. Figure 3.16 shows a quarter-wavelength monopole for mobile communication applications. This type of antenna normally uses a flexible antenna element and thus is called the quarter-

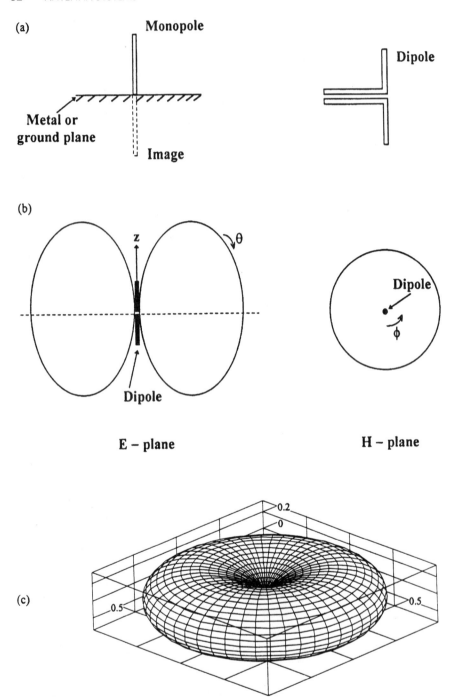

FIGURE 3.14   (*a*) Thin-wire dipole and monopole antenna. (*b*) *E*- and *H*-plane radiation patterns. (*c*) Three-dimensional view of the radiation pattern.

$\ell = 0.5\,\lambda_0$ $0.75\,\lambda_0$ λ_0 $1.25\,\lambda_0$ $1.5\,\lambda_0$ $1.75\,\lambda_0$ $2\,\lambda_0$ $2.25\,\lambda_0$ $2.75\,\lambda_0$

FIGURE 3.15 Radiation patterns of center-driven dipole assuming sinusoidal current distribution and very thin dipole. (From reference [3], with permission from McGraw-Hill.)

wavelength "whip" antenna. The antenna is mounted on the ground plane, which is the roof of a car. If the ground plane is assumed to be very large and made of a perfect conductor, the radiation patterns of this antenna would be the same as that of a half-wave dipole. However, the input impedance is only half that of a half-wave dipole.

There are various types of dipoles or monopoles used for wireless applications. A folded dipole is formed by joining two cylindrical dipoles at the ends, as shown in Fig. 3.17. The excitation of a folded dipole can be considered as a superposition of two modes, a symmetrical mode and an asymmetrical mode [6].

Figure 3.18 shows a sleeve antenna [7]. The coaxial cylindrical skirt behaves as a quarter-wavelength choke, preventing the antenna current from leaking into the outer surface of the coaxial cable. The choke on the lower part of the coaxial cable is used to improve the radiation pattern by further suppressing the current leakage. This antenna does not require a ground plane and has almost the same characteristics as that of a half-wavelength dipole antenna. The feeding structure is more suitable for vehicle mounting than the center-fed dipole antenna.

Other variations of monopole antennas are inverted L and inverted F antennas, as shown in Figs. 3.19a and b. These are low-profile antennas formed by bending a quarter-wavelength monopole element mounted on a ground plane into an L-shape or F-shape. The wire can be replaced by a planar element with wide-band characteristics [7]. The modified structure is shown in Fig. 3.19c. These types of antennas are used on hand-held portable telephones.

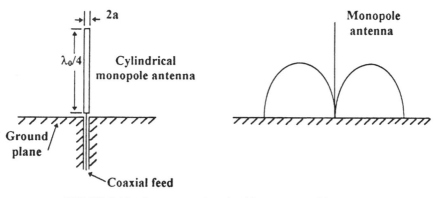

FIGURE 3.16 Quarter-wavelength whip antenna and its pattern.

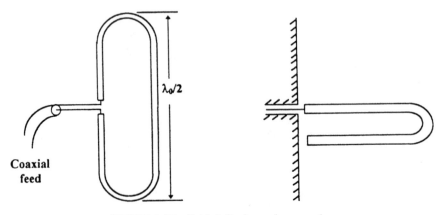

FIGURE 3.17 Folded dipoles and monopoles.

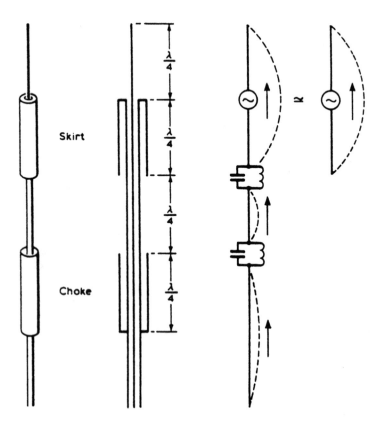

FIGURE 3.18 Sleeve antenna for vehicle application. (From reference [7], with permission from McGraw-Hill.)

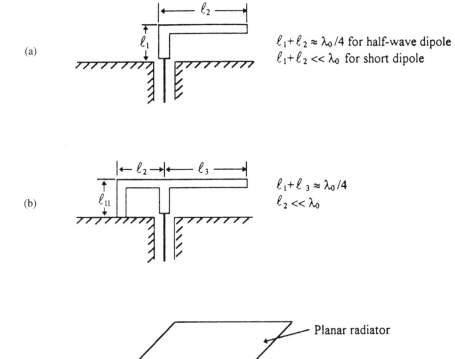

$\ell_1+\ell_2 \approx \lambda_0/4$ for half-wave dipole
$\ell_1+\ell_2 \ll \lambda_0$ for short dipole

$\ell_1+\ell_3 \approx \lambda_0/4$
$\ell_2 \ll \lambda_0$

Planar radiator

FIGURE 3.19   (*a*) Inverted *L*-antenna (ILA). (*b*) Inverted *F*-antenna (IFA). (*c*) Planar inverted *F*-antenna.

Example 3.2   An AM radio station operates at 600 kHz transmitting an output power of 100 kW using a monopole antenna as shown in Fig. 3.20. (a) What is the length of *l* if the antenna is an equivalent half-wave dipole? (b) What is the maximum rms electric field in volts per meter at a distance 100 km away from the station? The half-wave dipole has an antenna gain of 1.64.

Solution

(a)

$$\lambda_0 = \frac{c}{f} = \frac{3 \times 10^8 \text{ m/sec}}{600 \times 10^3 \text{ sec}^{-1}} = 500 \text{ m}$$

$$2l = \text{monopole and its image} = \tfrac{1}{2}\lambda_0$$

$$l = \tfrac{1}{4}\lambda_0 = 125 \text{ m}$$

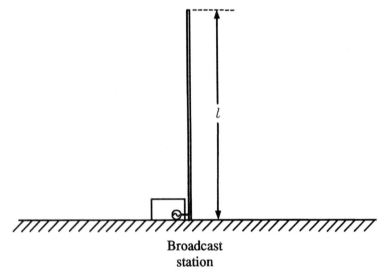

**Broadcast
station**

FIGURE 3.20   Broadcast station uses a monopole antenna.

(b) The power density for an isotropic antenna at a distance R is equal to $P_t/4\pi R^2$, as shown in Eq. (3.1). For a directive antenna with a gain G, the maximum power density is

$$P_{d,\text{max}} = G\frac{P_t}{4\pi R^2} = 1.64 \times \frac{100 \times 10^3 \text{ W}}{4 \times 3.14 \times (100 \times 10^3 \text{ m})^2}$$
$$= 1.31 \times 10^{-6} \text{ W/m}^2$$

The maximum power density occurs when $\theta = 90°$, that is, on the ground.
    From Eqs. (3.6) and (3.7),

$$P_d = \frac{E_{\text{rms}}^2}{\eta_0} \qquad \eta_0 = 377 \ \Omega$$

The maximum E_{rms} is

$$E_{\text{rms}} = \sqrt{P_{d,\text{max}}\eta_0} = 22.2 \text{ mV/m} \qquad \blacksquare$$

3.7   HORN ANTENNAS

The horn antenna is a transition between a waveguide and free space. A rectangular waveguide feed is used to connect to a rectangular waveguide horn, and a circular waveguide feed is for the circular waveguide horn. The horn antenna is commonly used as a feed to a parabolic dish antenna, a gain standard for antenna gain measurements, and as compact medium-gain antennas for various systems. Its gain can be calculated to within 0.1 dB accuracy from its known dimensions and is therefore used as a gain standard in antenna measurements.

For a rectangular pyramidal horn, shown in Fig. 3.21a, the dimensions of the horn for optimum gain can be designed by setting [3]

$$A = \sqrt{3\lambda_0 l_h}, \qquad B = \sqrt{2\lambda_0 l_e} \qquad (3.27)$$

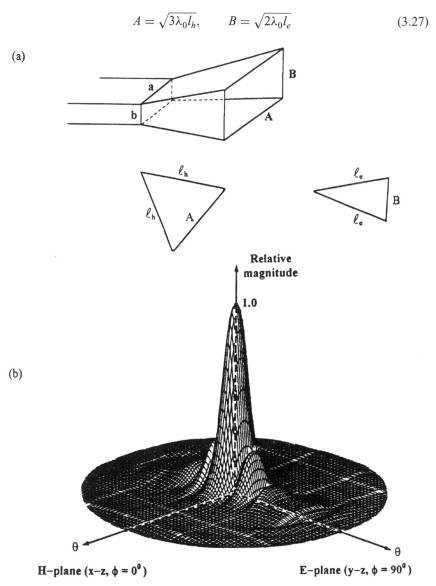

FIGURE 3.21 Rectangular pyramidal horn fed by a rectangular waveguide: (a) configuration and (b) typical three-dimensional radiation pattern.

where A and B are dimensions of the horn and l_e and l_h are the slant lengths of the horn, as shown in Fig. 3.21. The effective area is close to 50% of its aperture area, and its gain is given as

$$\text{Gain (in dB)} = 8.1 + 10 \log \frac{AB}{\lambda_0^2} \qquad (3.28)$$

As an example, a horn with $A = 9$ in. and $B = 4$ in. operating at 10 GHz will have a gain of 22.2 dB. A typical three-dimensional pattern is shown in Fig. 3.21b [8].

For an optimum-gain circular horn, shown in Fig. 3.22, the diameter should be designed as

$$D = \sqrt{3 l_c \lambda_0} \qquad (3.29a)$$

and

$$\text{Gain (in dB)} = 20 \log \frac{\pi D}{\lambda} - 2.82 \qquad (3.29b)$$

3.8 PARABOLIC DISH ANTENNAS

A parabolic dish is a high-gain antenna. It is the most commonly used reflector antenna for point-to-point satellites and wireless links.

A parabolic dish is basically a metal dish illuminated by a source at its focal point. The spherical wavefront illuminated by the source is converted into a planar wavefront by the dish, as shown in Fig. 3.23 [9].

For an illumination efficiency of 100%, the effective area equals the physical area:

$$A_e = \pi \left(\frac{D}{2} \right)^2 = A \qquad (3.30)$$

where D is the diameter of the dish.

In practice, the illumination efficiency η is typically between 55 and 75% due to the feed spillover, blockage, and losses. Using a 55% efficiency for the worst case, we have

$$A_e = \eta A = 0.55 \pi (\tfrac{1}{2} D)^2 \qquad (3.31)$$

and the gain of the antenna is, from (3.24),

$$G = \frac{4\pi}{\lambda_0^2} A_e = 0.55 \left(\frac{\pi D}{\lambda_0} \right)^2 \qquad (3.32)$$

The half-power beamwidth is approximately given by [from Eq. (3.25) with $K_1 = 70°$]

$$\text{HPBW} = 70 \frac{\lambda_0}{D} \quad \text{(deg)} \qquad (3.33)$$

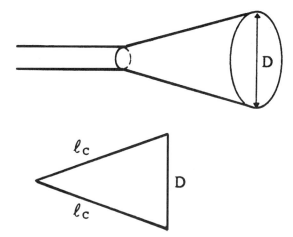

FIGURE 3.22   Circular horn fed by a circular waveguide.

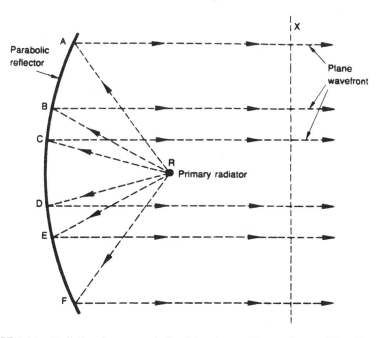

FIGURE 3.23   Radiation from a parabolic dish antenna. (From reference [9], with permission from Longman Scientific & Technical.)

Example 3.3 A parabolic dish antenna with a diameter of 3 ft is operated at 10 GHz. Determine the approximate gain, beamwidth, and distance for the far-field region operation. The illumination efficiency is 55%.

Solution

$$D = 3 \text{ ft} = 36 \text{ in.}$$

$$\lambda_0 = \frac{c}{f} = 3 \text{ cm} = 1.18 \text{ in.}$$

$$\text{Gain} = 10 \log\left[0.55 \left(\frac{\pi D}{\lambda_0} \right)^2 \right] = 37 \text{ dB or } 5047$$

$$\text{Beamwidth} = 70\frac{\lambda_0}{D} = 2.29°$$

For far-field region operation

$$R > \frac{2D^2}{\lambda_0} = 2196 \text{ in. or } 183 \text{ ft.} \qquad \blacksquare$$

It can be seen that the dish antenna provides a very high gain and narrow beam. The alignment of the dish antenna is usually very critical. The parabolic dish is generally fed by a horn antenna connected to a coaxial cable. There are four major feed methods: front feed, Cassegrain, Gregorian, and offset feed. Figure 3.24 shows these arrangements [9]. The front feed is the simplest method. The illumination efficiency is only 55–60%. The feed and its supporting structure produce aperture blockage and increase the sidelobe and cross-polarization levels. The Cassegrain method has the advantages that the feed is closer to other front-end hardware and a shorter connection line is needed. The Gregorian method is similar to the Cassegrain feed, but an elliptical subreflector is used. An illumination efficiency of 76% can be achieved. The offset feed method avoids the aperture blockage by the feed or subreflector. The sidelobe levels are smaller, and the overall size is smaller for the same gain.

At low microwave frequencies or ultrahigh frequencies (UHFs), a parabolic dish is big, so only a portion of the dish is used instead. This is called a truncated parabolic dish, commonly seen on ships. To make the dish lighter and to withstand strong wind, a dish made of metal mesh instead of solid metal can be used.

3.9   MICROSTRIP PATCH ANTENNAS

Microstrip patch antennas are widely used due to the fact that they are highly efficient, structurally compact, and conformal. One of the most common types of microstrip antenna is the rectangular patch.

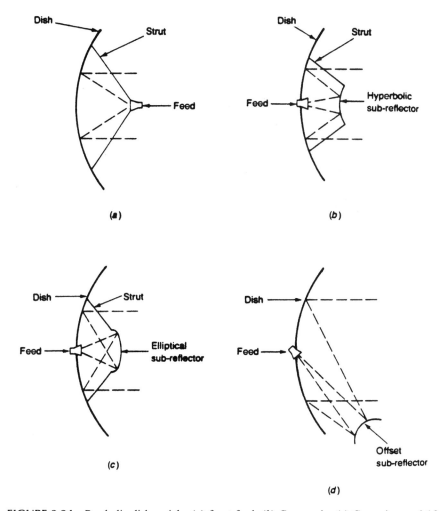

FIGURE 3.24   Parabolic dish aerials: (*a*) front feed; (*b*) Cassegrain; (*c*) Gregorian; and (*d*) offset-feed. (From reference [9], with permission from Longman Scientific & Technical.)

Figure 3.25 shows a typical rectangular patch antenna with width W and length L over a grounded dielectric plane with dielectric constant ε_r. Ideally, the ground plane on the underside of the substrate is of infinite extent. Normally, the thickness of the dielectric substrate, h, is designed to be $\leq 0.02\lambda_g$, where λ_g is the wavelength in the dielectric.

There are several theories that can be used for the analysis and design of microstrip patch antennas. The first one is the transmission line model [10]. This theory is based on the fact that the rectangular patch is simply a very wide transmission line terminated by the radiation impedance. The transmission line model can predict properties only approximately, because the accepted formulas that

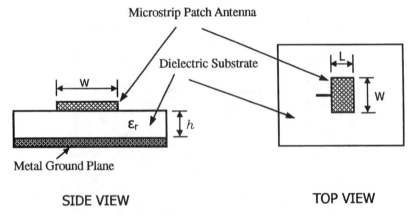

FIGURE 3.25 Rectangular patch antenna.

describe microstrip transmission line characteristics are either approximations or empirically fit to measured data. The second method is the cavity model [11]. This model assumes the rectangular patch to be essentially a closed resonant cavity with magnetic walls. The cavity model can predict all properties of the antenna with high accuracy but at the expense of much more computation effort than the transmission line model.

The patch antenna can be approximately designed by using the transmission line model. It can be seen from Fig. 3.26 that the rectangular patch with length L and width W can be viewed as a very wide transmission line that is transversely resonating, with the electric field varying sinusoidally under the patch along its resonant length. The electric field is assumed to be invariant along the width W of the patch. Furthermore, it is assumed that the antenna's radiation comes from fields leaking out along the width, or radiating edges, of the antenna.

The radiating edges of the patch can be thought of as radiating slots connected to each other by a microstrip transmission line. The radiation conductance for a single slot is given as

$$G = \frac{W^2}{90\lambda_0^2} \quad \text{for } W < \lambda_0 \tag{3.34a}$$

$$G = \frac{W}{120\lambda_0} \quad \text{for } W > \lambda_0 \tag{3.34b}$$

Similarly, the radiation susceptance of a single slot [12] is given as

$$B = \frac{k_0 \, \Delta l \sqrt{\varepsilon_{\text{eff}}}}{Z_0} \tag{3.35a}$$

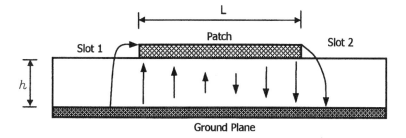

RADIATING PATCH ANTENNA

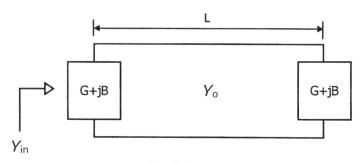

EQUIVALENT CIRCUIT

FIGURE 3.26 Transmission line model of a patch antenna.

where

$$Z_0 = \frac{120\pi h}{W\sqrt{\varepsilon_{\text{eff}}}} \tag{3.35b}$$

with

$$\varepsilon_{\text{eff}} = \frac{\varepsilon_r + 1}{2} + \frac{\varepsilon_r - 1}{2}\left(1 + \frac{12h}{W}\right)^{-1/2} \tag{3.35c}$$

$$\Delta l = 0.412h\left(\frac{\varepsilon_{\text{eff}} + 0.3}{\varepsilon_{\text{eff}} - 0.258}\right)\frac{(W/h) + 0.264}{(W/h) + 0.8} \tag{3.35d}$$

where $k_0 = 2\pi/\lambda_0$ is the free-space wavenumber, Z_0 is the characteristic impedance of the microstrip line with width W, ε_{eff} is referred to as the effective dielectric constant, and Δl is the correction term called the edge extension, accounting for the fringe capacitance. In Fig. 3.26 it can be seen that the fields slightly overlap the edges of the patch, making the electrical length of the patch slightly larger than its physical length, thus making the edge extension necessary.

To determine the radiation impedance of the antenna, we combine the slot impedance with the transmission line theory. The admittance of a single slot is given in Eqs. (3.34) and (3.35). The microstrip patch antenna is merely two slots in parallel separated by a transmission line with length L, which has a characteristic admittance Y_0. The input admittance at the radiating edge can be found by adding the slot admittance to the admittance of the second slot by transforming it across the length of the patch using the transmission line equation. The result is

$$Y_{in} = Y_{slot} + Y_0 \frac{Y_{slot} + jY_0 \tan \beta(L + 2\,\Delta l)}{Y_0 + jY_{slot} \tan \beta(L + 2\,\Delta l)} \tag{3.36}$$

where $\beta = 2\pi\sqrt{\varepsilon_{eff}}/\lambda_0$ is the propagation constant of the microstrip transmission line. At resonance ($L + 2\,\Delta l = \frac{1}{2}\lambda_g$), this reduces to two slots in parallel, giving an input admittance twice that of Eq. (3.34),

$$Y_{in} = 2G \tag{3.37}$$

More generally, the input admittance at a point inside the patch at a given distance y_1 from the radiating edge can be found by using the transmission line equation to transform the slot admittance across the patch by a distance y_1 [13]. We add this to the admittance from the other slot, which is transformed a distance $y_2 = (L + 2\,\Delta l) - y_1$ so the two admittances are at the same point. This result gives

$$Y_{in} = Y_0 \frac{Y_{slot} + jY_0 \tan \beta y_1}{Y_0 + jY_{slot} \tan \beta y_1} + Y_0 \frac{Y_{slot} + jY_0 \tan \beta y_2}{Y_0 + jY_{slot} \tan \beta y_2} \tag{3.38}$$

The patch antenna resonates when the imaginary part of (3.38) disappears.

Perhaps a more intuitive picture of resonance can be seen from field distribution under the patch in Fig. 3.26. In order to resonate, the effective length [adding twice the length extension found in Eq. (3.35d) onto the physical length] must be equal to half a transmission line wavelength. In other words,

$$(L + 2\,\Delta l) = \frac{\lambda_g}{2} = \frac{\lambda_0}{2\sqrt{\varepsilon_{eff}}} \tag{3.39}$$

from which we can determine the resonant frequency (which is the operating frequency) in terms of patch dimensions:

$$f_r = \frac{c}{2\sqrt{\varepsilon_{eff}}(L + 2\,\Delta l)} \tag{3.40}$$

Here, W is not critical but can be selected as

$$W = \frac{c}{2f_r} \left(\frac{\varepsilon_r + 1}{2} \right)^{-1/2} \tag{3.41}$$

From Eqs. (3.41), (3.35c), (3.35d), and (3.40), one can determine W and L if ε_r, h, and f_r are known. Equations (3.34) and (3.37) give the input admittance at the resonant or operating frequency.

Example 3.4   Design a microstrip patch antenna operating at 3 GHz. The substrate is Duroid 5880 ($\varepsilon_r = 2.2$) with a thickness of 0.030 in. The antenna is fed by a 50-Ω line, and a quarter-wavelength transformer is used for impedance matching, as shown in Fig. 3.27.

Solution

$$f_r = 3 \text{ GHz}, h = 0.030 \text{ in.} = 0.0762 \text{ cm}$$
$$\lambda_0 = \frac{c}{f_r} = 10 \text{ cm}$$

From Eq. (3.41)

$$W = \frac{c}{2f_r} \left(\frac{\varepsilon_r + 1}{2} \right)^{-1/2} = 3.95 \text{ cm}$$

From Eq. (3.35c)

$$\varepsilon_{\text{eff}} = \frac{\varepsilon_r + 1}{2} + \frac{\varepsilon_r - 1}{2} \left[1 + \frac{12h}{W} \right]^{-1/2} = 2.14$$

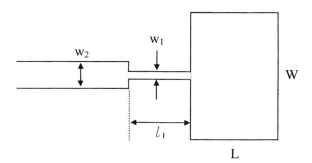

FIGURE 3.27   Design of a rectangular patch antenna.

From Eq. (3.35d)

$$\Delta l = 0.412h \left(\frac{\varepsilon_{\text{eff}} + 0.3}{\varepsilon_{\text{eff}} - 0.258} \right) \left(\frac{W/h + 0.264}{W/h + 0.8} \right) = 0.04 \text{ cm}$$

From Eq. (3.40)

$$L = \frac{c}{2f_r\sqrt{\varepsilon_{\text{eff}}}} - 2\,\Delta l = 3.34 \text{ cm}$$

Since $W < \lambda_0$, $G = W^2/90\lambda_0^2$ from (3.34),

$$Y_{\text{in}} = 2G = \frac{1}{45} \frac{W^2}{\lambda_0^2} = \frac{1}{R_{\text{in}}}$$

$$R_{\text{in}} = 288 \ \Omega = \text{input impedance}$$

The characteristic impedance of the transformer is

$$Z_{0T} = \sqrt{R_{\text{in}} \times 50} = 120 \ \Omega$$

From Figs. 2.24 and 2.25,

$$\frac{W_1}{h} \approx 0.58 \quad \text{and} \quad \frac{\lambda_g\sqrt{\varepsilon_r}}{\lambda_0} \approx 1.13$$

Therefore, the impedance transformer has a width W_1 and length l_1 given by

$$W_1 = 0.0442 \text{ cm} \qquad l_1 = \tfrac{1}{4}\lambda_g = 1.90 \text{ cm}$$

For a 50-Ω line W_2 can be estimated from Fig. 2.24 and found to be 0.228 cm.

Similar results can be obtained by using Eqs. (2.84) and (2.86); we have $W_1 = 0.044$ cm, $\varepsilon_{\text{eff}} = 1.736$, and $l_1 = 1.89$ cm. For the 50-Ω line, $W_2 = 0.238$ cm from Eq. (2.84) ∎

Patch antennas can be fed by many different ways [13–16]. The most common feed methods are the coaxial probe feed and microstrip line edge feed, as shown in Fig. 3.28. The probe feed method is simple but is not attractive from the fabrication point of view. The edge feed method has the advantage that both patch and feed line can be printed together. Other feed methods are electromagnetically coupled microstrip line feed, aperture-coupling feed, slot line feed, coplanar waveguide feed, and so on.

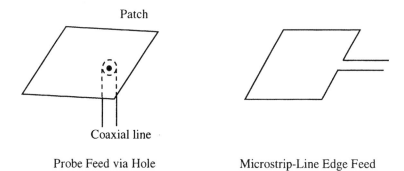

Probe Feed via Hole   Microstrip-Line Edge Feed

FIGURE 3.28   Microstrip patch antenna feed methods.

The radiation patterns of a microstrip patch antenna can be calculated based on electromagnetic analysis [13–16]. Typical radiation patterns are shown in Fig. 3.29 [17]. Typical half-power beamwidth is 50°–60°, and typical gain ranges from 5 to 8 dB.

In many wireless applications, CP antennas are required. A single square patch can support two degenerate modes at the same frequency with the radiated fields linearly polarized in orthogonal directions. Circular polarization can be accomplished by using a proper feed network with a 90° hybrid coupler, as shown in Fig. 3.30a. Other methods can also be employed to obtain circular polarization by perturbing the patch dimensions without the need of the hybrid coupler feed network. Some of these methods are shown in Fig. 3.30b.

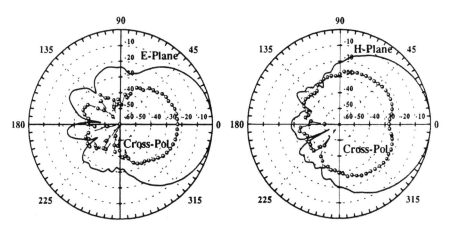

FIGURE 3.29   Radiation patterns of an inverted microstrip patch antenna. (From reference [17], with permission from IEEE.)

(a)

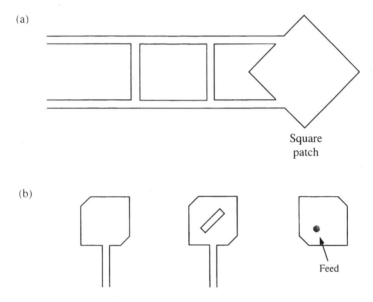

Square
patch

(b)

Feed

FIGURE 3.30   Circularly polarized square patch antennas.

3.10   ANTENNA ARRAYS AND PHASED ARRAYS

Single antennas are often limited for many applications because of a large HPBW and, consequently, a lower gain. For many applications, a high-gain, narrow pencil beam is required. Since most antennas have dimensions that are on the order of one wavelength, and since beamwidth is inversely proportional to antenna size, more than one antenna is required to sharpen the radiation beam. An array of antennas working simultaneously can focus the reception or transmission of energy in a particular direction, which increases the useful range of a system.

Considering the one-dimensional linear array shown in Fig. 3.31, the radiated field from a set of sources can be given by

$$E_{\text{total}} = I_1 f_1(\theta, \phi) \rho_1 \frac{e^{-j(k_0 r_1 - \Phi_1)}}{4\pi r_1} + I_2 f_2(\theta, \phi) \rho_2 \frac{e^{-j(k_0 r_2 - \Phi_2)}}{4\pi r_2} + \cdots$$

$$+ I_i f_i(\theta, \phi) \rho_i \frac{e^{-j(k_0 r_i - \Phi_i)}}{4\pi r_i} + \cdots \tag{3.42}$$

where I_i, ρ_i, and ϕ_i are the ith element's magnitude, polarization, and phase, respectively; $f_i(\theta, \phi)$ is the radiation pattern of the ith element and r_i is the distance from the ith element to an arbitrary point in space; and k_0 is the propagation constant, equal to $2\pi/\lambda_0$.

Far Field Amplitude Variations

$$\text{Factor} = \sum_{i=1}^{N} I_i e^{j(i-1)(kd\cos\theta-\Phi)}$$

ar array along z axis.

is aligned for copolarization (i.e.,
form spacing d. It is oriented along
element is placed at the origin, and
d using the following:

$$(3.43)$$

$d \cos \theta$

e given by

$-j(i-1)(k_0 d \cos\theta - \Phi)$

array factor $\qquad(3.44)$

equations above is made up of an
ray factor (AF). This is known as

e AF can be formulated without
array. For the sake of convenience
radiators and $I_i = 1$. For an array
given as

$$AF = 1 + e^{-j(k_0 d \cos\theta - \Phi)} + e^{-j2(k_0 d \cos\theta - \Phi)} + \cdots + e^{-j(N-1)(k_0 d \cos\theta - \Phi)} \qquad (3.45a)$$

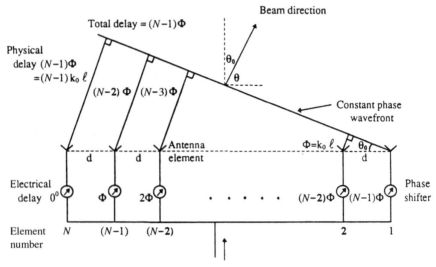

FIGURE 3.32 An N-element linear array with progressively larger phase delay from left to right.

or

$$AF = \sum_{n=0}^{N-1} e^{-jn\psi} \tag{3.45b}$$

where $\psi = k_0 d \cos \theta - \Phi$.

The parameter Φ is the progressive phase shift across the array, which means that there is a phase difference of Φ between the currents on adjacent elements. The progressive phase shift causes the radiation emitted from the array to have a constant phase front that is pointing at the angle θ_0 (where $\theta_0 = 90° - \theta$), as seen in Fig. 3.32. By varying the progressive phase shift across the array, the constant phase front is also varied.

To see how this is accomplished, note that the array factor in (3.45b) is a maximum when the exponential term equals 1. This happens when $\psi = 0$ or when

$$\Phi = k_0 d \, \sin(\theta_0) \tag{3.46a}$$

or

$$\theta_0 = \text{scanning angle} = \sin^{-1}\left(\frac{\Phi}{k_0 d}\right) \tag{3.46b}$$

As Φ is varied, it must satisfy (3.46) in order to direct the constant phase front of the radiation at the desired angle θ_0 (scanning the main beam). This is the basic concept used in a phased array.

Alternatively, Eq. (3.46) can be derived by examining Fig. 3.32. For each element, at the constant phase wavefront, the total phase delay should be the same for all elements. The total phase delay equals the summation of the electrical phase delay due to the phase shifter and the physical phase delay. From any two neighboring elements, elements 1 and 2 for example, we have

$$l = d \sin \theta_0 \tag{3.47}$$

Therefore,

$$\Phi = k_0 l = k_0 d \sin \theta_0 \tag{3.48}$$

Equation (3.48) is the same as (3.46a).

For an array antenna, the beamwidth and gain can be estimated from the number of elements. With the elements spaced by half-wavelengths to avoid the generation of grating lobes (multiple beams), the number of radiating elements N for a pencil beam is related to the half-power (or 3-dB) beamwidth by [18]

$$N \approx \frac{10,000}{(\theta_{BW})^2} \tag{3.49}$$

where θ_{BW} is the half-power beamwidth in degrees. From (3.49), we have

$$\theta_{BW} \approx \frac{100}{\sqrt{N}}$$

The corresponding antenna array gain is

$$G \approx \eta \pi N \tag{3.50}$$

where η is the aperture efficiency.

In a phased array, the phase of each antenna element is electronically controllable. One can change the phase of each element to make the array electronically steerable. The radiation beam will point to the direction that is normal to the constant phase front. This front is adjusted electronically by individual control of the phase of each element. In contrast to the mechanically steerable beam, the beam in a phased array can be steered much faster, and the antenna array is physically stationary.

The array factor is a periodic function. Hence, it is possible to have a constant phase front in several directions, called grating lobes. This can happen when the argument in the exponential in Eq. (3.45b) is equal to a multiple of 2π. To scan to a given angle, θ_0, as in Fig. 3.32, Φ must be chosen to satisfy $\Phi = k_0 d \sin(\theta_0)$ as before. Thus, $\psi = -2\pi = k_0 d(\cos \theta - \sin \theta_0)$. For the given scan direction, a grating lobe will begin to appear in the end-fire direction ($\theta = 180°$) when $-k_0 d[1 + \sin\theta_0] = -2\pi$. Dividing out 2π from the equation, we come up with the condition that

$$\frac{d}{\lambda_0} = \frac{1}{1 + \sin\theta_0} \tag{3.51}$$

Grating lobes reduce the array's ability to focus the radiation in a specific area of angular space (directivity) and are undesirable in the antenna pattern. The spacing between adjacent elements should be less than the distance defined in Eq. (3.51) to avoid grating lobes.

By adjusting the phase and spacing between elements in two dimensions, we can extend this theory to a two-dimensional array. Such an array would make scanning possible in two perpendicular directions. The array factor in (3.45b) is modified to

$$AF = \sum_{m=0}^{M-1} e^{-jm(k_0 d_x \cos\theta \cos\phi - \Phi_x)} \sum_{n=0}^{N-1} e^{-jn(k_0 d_y \cos\theta \sin\phi - \Phi_y)} \tag{3.52}$$

where d_x and d_y define the element spacing in the x and y directions, respectively, and Φ_x and Φ_y are the progressive phase shifts in the x and y directions, respectively.

The arrays can be fed by two major methods: corporate feed method and traveling-wave feed method, as shown in Fig. 3.33. In the corporate feed method, 3-dB power dividers are used to split the input power and deliver it to each element.

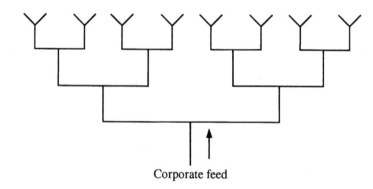

Corporate feed

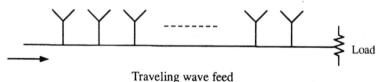

Traveling wave feed

FIGURE 3.33 Antenna array feeding methods.

Each power splitter introduces some losses. In the traveling-wave feed method, antenna elements are coupled to a transmission line. The power coupled to each element is controlled by the coupling mechanisms.

Figure 3.34 shows a 16 × 16 microstrip patch antenna array with 256 elements using a corporate feed. A coaxial feed is connected to the center of the array from the other side of the substrate. Power dividers and impedance-matching sections are used to couple the power to each element for radiation.

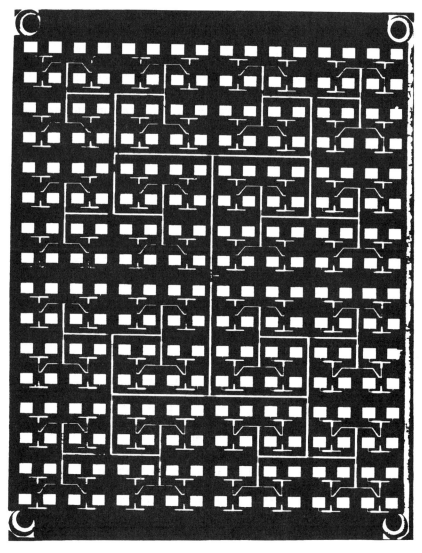

FIGURE 3.34 A 16 × 16 microstrip patch antenna array. (Courtesy of Omni-Patch Designs.)

3.11 ANTENNA MEASUREMENTS

Antennas can be measured in an indoor antenna chamber or an outdoor antenna range. Figure 3.35 shows a typical setup for the antenna testing [2]. The antenna under test (AUT) is located at the far-field region on top of a rotating table or positioner. A standard gain horn is used as a transmitting antenna, and the AUT is normally used as a receiving antenna. For the indoor range, the setup is placed inside an anechoic chamber. The chamber walls are covered with wave-absorbing material to isolate the AUT from the building structure and simulate the free-space unbounded medium.

The system is first calibrated by two standard gain horns. Parameters to be measured include antenna gain, antenna patterns, sidelobes, half-power beamwidth, directivity, cross-polarization, and back radiation. In most cases, external objects, finite ground planes, and other irregularities change the radiation patterns and limit the measurement accuracies.

Figure 3.36 shows a standard anechoic chamber for antenna measurement for far-field pattern measurements. Near-field and compact indoor ranges are also available for some special purposes.

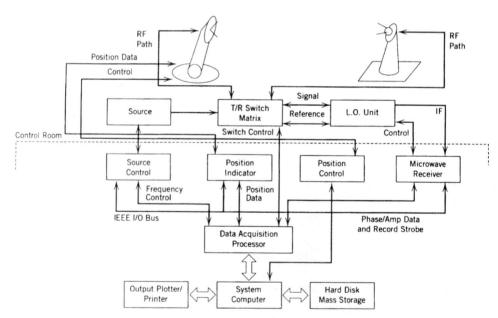

FIGURE 3.35 Typical antlenna range setup [2].

FIGURE 3.36   Standard indoor anechoic chamber at Texas A&M University.

PROBLEMS

3.1   Design a half-wave dipole at 1 GHz. Determine (a) the length of this dipole, (b) the length of the monopole if a monopole is used instead of a dipole, and (c) the power density (in millwatts per square centimeters) and rms E-field (in volts per meter) at a distance of 1 km if the transmit power $P_t = 10$ W.

3.2   At 2 GHz, what is the length in centimeters of a half-wavelength dipole? What is the length of the equivalent monopole?

3.3   An FM radio station operating at 10 MHz transmits 1000 W using a monopole antenna, as shown in Fig. P3.3.

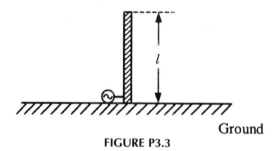

Ground

FIGURE P3.3

(a) Design l in meters such that the antenna is an equivalent half-wave dipole.

(b) The half-wave dipole has a gain of 1.64. What is the maximum rms electric field (in volts per meter) at a distance of 100 km away from the station?

3.4 A transmitter has a power output of 100 W and uses a dipole with a gain of 1.5. What is the maximum rms electric field at a distance of 1 km away from the transmitter?

3.5 A rectangular horn antenna has $A = 6$ in. and $B = 4$ in. operating at 12 GHz. Calculate (a) the antenna gain in decibels and (b) the maximum rms E-field at a distance of 100 m from the antenna if the transmit power is 100 W (E is in volts per meter).

3.6 Calculate the gain (in decibels) and 3-dB beamwidth (in degrees) for a dish antenna with the following diameters at 10 GHz: (a) 5 ft and (b) 10 ft. Determine the far-field zones for both cases ($\eta = 55\%$).

3.7 At 10 GHz, design a dish antenna with a gain of 60 dB. What is the diameter of this antenna? What is the beamwidth in degrees? Note that this is a very high gain antenna ($\eta = 55\%$).

3.8 A dish antenna assuming a 55% illumination efficiency with a diameter of 3 m is attached to a transmitter with an output power of 100 kW. The operating frequency is 20 GHz. (a) Determine the maximum power density and electric (rms) field strength at a distance of 10 km away from this antenna. (b) Is this distance located at the far-field zone? (c) Is it safe to have someone inside the beam at this distance? (Note that the U.S. safety standard requires $P_{d,max} < 10$ mW/cm^2.)

3.9 A high-power radar uses a dish antenna with a diameter of 4 m operating at 3 GHz. The transmitter produces 200 kW CW power. What is the minimum safe distance for someone accidentally getting into the main beam? (The U.S. standard requires power density < 10 mW/cm^2 for safety.)

3.10 A parabolic dish antenna has a diameter of 1 m operating at 10 GHz. The antenna efficiency is assumed to be 55%. (a) Calculate the antenna gain in decibels. (b) What is the 3-dB beamwidth in degrees? (c) What is the maximum power density in watts per square meter at a distance of 100 m away from the antenna? The antenna transmits 10 W. (d) What is the power density at 1.05° away from the peak?

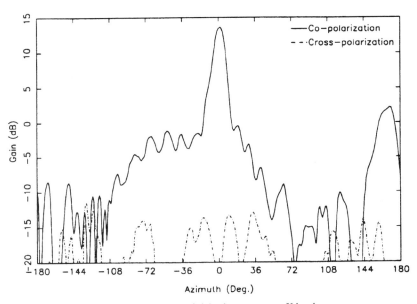

E-plane radiation pattern of eight element array at X-band

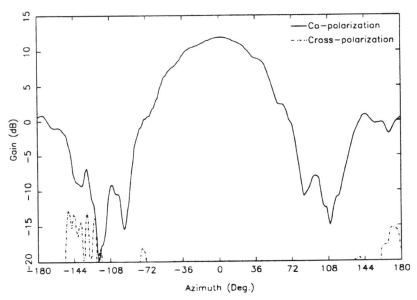

H-plane radiation pattern of eight element array at X-band

FIGURE P3.13

3.11 A parabolic dish antenna has a diameter of 2 m. The antenna is operating at 10 GHz with an efficiency of 55%. The first sidelobe is 20 dB below the main-beam peak. (a) Calculate the antenna gain in decibels and the antenna's effective area in square meters. (b) Calculate the maximum power density in watts per square meter and E-field in volts per meter at a distance of 1 km away from the antenna. The power transmitted by the antenna is 100 W. (c) Compute the half-power beamwidth in degrees. (d) Calculate the power density in watts per square meter at the first-sidelobe location.

3.12 A direct-TV parabolic dish antenna has a diameter of 12 in. operating at 18 GHz with an efficiency of 60%. Calculate (a) the antenna gain in decibels, (b) the effective area in square centimeters, and (c) the half-power beamwidth in degrees. (d) Sketch the antenna pattern.

3.13 The E- and H-plane patterns shown in Fig. P3.13 are obtained from a dielectric waveguide fed microstrip patch array [4]. The array consists of 1×8 elements and operates at 10 GHz. Determine (a) the beamwidths for both patterns, (b) the maximum sidelobe levels for both patterns, (c) the backscattering levels for both patterns, and (d) the cross-polarization levels for both patterns.

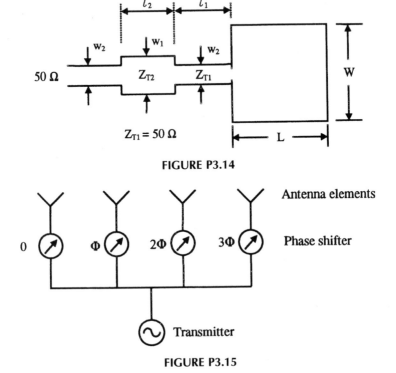

FIGURE P3.14

FIGURE P3.15

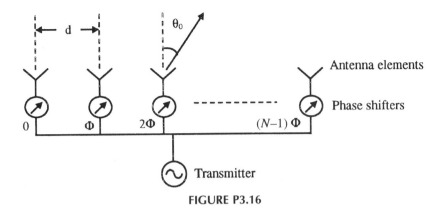

FIGURE P3.16

3.14 In the design of a matching circuit for a patch antenna in Example 3.4, since W_1 is small and a narrow microstrip line is lossy, an alternative design is to use a quarter-wavelength 50-Ω line followed by a quarter-wavelength low-impedance line as shown in Fig. P3.14. Determine the dimensions for W_1, W_2, l_1 and l_2 (see figure).

3.15 A four-element antenna array is shown in Fig. P3.15 with a phase shifter on each element. The separation between neighboring elements is $\lambda_0/2$. What is the scan angle θ_0 when Φ is equal to 90 degrees?

3.16 An N-element phased array has an antenna element separation d of $0.75\lambda_0$ (see Fig. P3.16). What are the scan angles when the progressive phase shift Φ equals 45 degrees and 90 degrees?

REFERENCES

1. R. E. Collin, *Foundations for Microwave Engineering*, McGraw-Hill, New York, 1st ed., 1966, 2nd ed., 1992.

2. J. A. Navarro and K. Chang, *Integrated Active Antennas and Spatial Power Combining*, John Wiley & Sons, New York, 1996.

3. R. C. Johnson, Ed., *Antenna Engineering Handbook*, 3rd ed., McGraw-Hill, New York, 1993.

4. S. Kanamaluru, M. Li, and K. Chang, "Analysis and Design of Aperture-Coupled Microstrip Patch Antennas and Arrays Fed by Dielectric Image Line," *IEEE Trans. Antennas Propagat.*, Vol. AP-44, pp. 964–974 1996.

5. L. C. Shen and J. A. Kong, *Applied Electromagnetism*, 2nd ed., PWS Engineering, Boston, MA, 1987.

6. S. Uda and Y. Mushiake, *Yagi-Uda Antenna*, Marnzen Co., Tokyo, 1954, p. 19.

7. H. Jasik, Ed., *Antenna Engineering Handbook*, McGraw-Hill, New York, 1961, pp. 22–25.

8. C. A. Balanis, *Antenna Theory*, 2nd ed., John Wiley & Sons, New York, 1997.

9. D. C. Green, *Radio Systems Technology*, Longman Scientific & Technical, Essex, United Kingdom, 1990.

10. K. R. Carver and J. W. Mink, "Microstrip Antenna Technology," *IEEE Trans. Antennas Propagat.*, Vol. AP-29, pp. 2–24, 1981.

11. Y. T. Lo, D. Solomon, and W. F. Richards, "Theory and Experiment on Microstrip Antennas," *IEEE Trans. Antennas Propagat.*, Vol. AP-27, pp. 137–145 1979.

12. R. E. Munson, "Conformal Microstrip Antennas and Microstrip Phased Arrays," *IEEE Trans. Antennas Propagat.*, Vol. AP-22, pp. 74–78, 1974.

13. K. Chang, Ed., *Handbook of Microwave and Optical Components*, Vol. 1, John Wiley & Sons, New York, 1990.

14. J. R. James, P. S. Hall, and C. Wood, *Microstrip Antenna Theory and Design*, Peter Peregrinus, Stevanage, United Kingdom, pp. 79–80, 1981.

15. I. J. Bahl and P. Bhartia, *Microstrip Antennas*, Artech House, Norwood, MA, 1980.

16. J. R. James and P. S. Hall, *Handbook of Microstrip Antennas*, Peter Peregrinus, London, 1989.

17. J. A. Navarro, J. McSpadden, and K. Chang, "Experimental Study of Inverted Microstrip for Integrated Antenna Applications," *1994 IEEE-AP International Antennas and Propagation Symposium Digest*, Seattle, WA, 1994, pp. 920–923.

18. M. Skolnik, Ed., *Radar Handbook*, 2nd ed., McGraw-Hill, New York, 1990, Ch. 7.

CHAPTER FOUR

Various Components and Their System Parameters

4.1 INTRODUCTION AND HISTORY

An RF and microwave system consists of many different components connected by transmission lines. In general, the components are classified as passive components and active (or solid-state) components. The passive components include resistors, capacitors, inductors, connectors, transitions, transformers, tapers, tuners, matching networks, couplers, hybrids, power dividers/combiners, baluns, resonators, filters, multiplexers, isolators, circulators, delay lines, and antennas. The solid-state devices include detectors, mixers, switches, phase shifters, modulators, oscillators, and amplifiers. Strictly speaking, active components are devices that have negative resistance capable of generating RF power from the DC biases. But a more general definition includes all solid-state devices.

Historically, wires, waveguides, and tubes were commonly used before 1950. After 1950, solid-state devices and integrated circuits began emerging. Today, monolithic integrated circuits (or chips) are widely used for many commercial and military systems. Figure 4.1 shows a brief history of microwave technologies. The commonly used solid-state devices are MESFETs (metal–semiconductor field-effect transistors), HEMTs (high-electron-mobility transistors), and HBTs (heterojunction bipolar transistors). Gallium–arsenide semiconductor materials are commonly used to fabricate these devices and the MMICs, since the electron mobility in GaAs is higher than that in silicon. Higher electron mobility means that the device can operate at higher frequencies or higher speeds. Below 2 GHz, silicon technology is dominant because of its lower cost and higher yield. The solid-state devices used in RF are mainly silicon transistors, metal–oxide–semiconductor FETs (MOSFETs), and complementary MOS (CMOS) devices. High-level monolithic integration in chips is widely used for RF and low microwave frequencies.

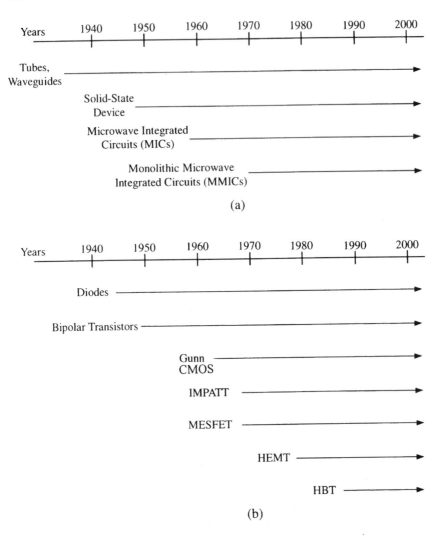

FIGURE 4.1 History of microwave techniques: (*a*) technology advancements; (*b*) solid-state devices.

In this chapter, various components and their system parameters will be discussed. These components can be represented by the symbols shown in Fig. 4.2. The design and detailed operating theory will not be covered here and can be found in many other books [1–4]. Some components (e.g., antennas, lumped *R*, *L*, *C* elements, and matching circuits) have been described in Chapters 2 and 3 and will not be repeated here. Modulators will be discussed in Chapter 9.

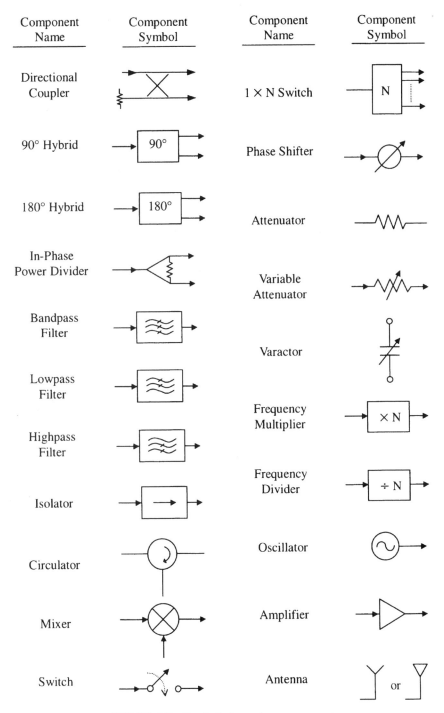

FIGURE 4.2 Symbols for various components.

4.2   COUPLERS, HYBRIDS, AND POWER DIVIDERS/COMBINERS

Couplers and hybrids are components used in systems to combine or divide signals. They are commonly used in antenna feeds, frequency discriminators, balanced mixers, modulators, balanced amplifiers, phase shifters, monopulse comparators, automatic signal level control, signal monitoring, and many other applications. A good coupler or hybrid should have a good VSWR, low insertion loss, good isolation and directivity, and constant coupling over a wide bandwidth.

A directional coupler is a four-port device with the property that a wave incident in port 1 couples power into ports 2 and 3 but not into 4, as shown in Fig. 4.3 [5]. The structure has four ports: input, direct (through), coupled, and isolated. The power P_1 is fed into port 1, which is matched to the generator impedance; P_2, P_3, and P_4 are the power levels available at ports 2, 3, and 4, respectively. The three important parameters describing the performance of the coupler are coupling factor, directivity, and isolation, defined by

$$\text{Coupling factor (in dB): } C = 10 \log \frac{P_1}{P_3} \tag{4.1}$$

$$\text{Directivity (in dB): } D = 10 \log \frac{P_3}{P_4} \tag{4.2}$$

$$\text{Isolation (in dB): } I = 10 \log \frac{P_1}{P_4}$$

$$= 10 \log \frac{P_1}{P_3} \frac{P_3}{P_4} = 10 \log \frac{P_1}{P_3} + 10 \log \frac{P_3}{P_4}$$

$$= C + D \tag{4.3}$$

In general, the performance of the coupler is specified by its coupling factor, directivity, and terminating impedance. The isolated port is usually terminated by a matched load. Low insertion loss and high directivity are desired features of the coupler. Multisection designs are normally used to increase the bandwidth.

Example 4.1   A 10-dB directional coupler has a directivity of 40 dB. If the input power $P_1 = 10$ mW, what are the power outputs at ports 2, 3, and 4? Assume that the coupler (a) is lossless and (b) has an insertion of 0.5 dB.

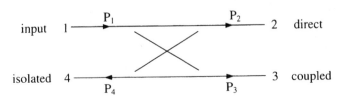

FIGURE 4.3   Directional coupler.

Solution   (a) For a lossless case, C (dB) = 10 dB = $10 \log(P_1/P_3) = P_1$ (dB) − P_3 (dB):

$$P_1 = 10 \text{ mW} = 10 \text{ dBm}$$

$$P_3 = P_1 - C = 10 \text{ dBm} - 10 \text{ dB} = 0 \text{ dBm} = 1 \text{ mW}$$

$$D \text{ (dB)} = 40 \text{ dB} = 10 \log \frac{P_3}{P_4} = P_3 \text{ (dB)} - P_4 \text{ (dB)}$$

$$P_4 = P_3 \text{ (dB)} - D \text{ (dB)} = 0 \text{ dBm} - 40 \text{ dB} = -40 \text{ dBm}$$

$$= 0.0001 \text{ mW}$$

$$P_2 = P_1 - P_3 - P_4 \approx 9 \text{ mW or } 9.5 \text{ dBm}$$

(b) For the insertion loss of 0.5 dB, let us assume that this insertion loss is equal for all three ports:

$$\text{Insertion loss} = \text{IL} = \alpha_L = 0.5 \text{ dB}$$

$$P_3 = 0 \text{ dBm} - 0.5 \text{ dB} = -0.5 \text{ dBm} = 0.89 \text{ mW}$$

$$P_4 = -40 \text{ dBm} - 0.5 \text{ dB} = -40.5 \text{ dBm} = 0.000089 \text{ mW}$$

$$P_2 = 9.5 \text{ dBm} - 0.5 \text{ dB} = 9 \text{ dBm} = 7.9 \text{ mW} \qquad \blacksquare$$

Hybrids or hybrid couplers are commonly used as 3-dB couplers, although some other coupling factors can also be achieved. Figure 4.4 shows a 90° hybrid. For the 3-dB hybrid, the input signal at port 1 is split equally into two output signals at ports 2 and 3. Ports 1 and 4 are isolated from each other. The two output signals are 90° out of phase. In a microstrip circuit, the hybrid can be realized in a branch-line type of circuit as shown in Fig. 4.4. Each arm is $\frac{1}{4}\lambda_g$ long. For a 3-dB coupling, the characteristic impedances of the shunt and series arms are: $Z_p = Z_0$ and $Z_s = Z_0/\sqrt{2}$, respectively, for optimum performance of the coupler [2, 3, 5]. The characteristic impedance of the input and output ports, Z_0, is normally equal to 50 Ω for a microstrip line. The impedances of the shunt and series arms can be designed to other values for different coupling factors [5]. It should be mentioned that port 4 can also be used as the input port; then port 1 becomes the isolated port due to the symmetry of the circuit. The signal from port 4 is split into two output signals at ports 2 and 3.

The 180° hybrid has characteristics similar to the 90° hybrid except that the two output signals are 180° out of phase. As shown in Fig. 4.5, a hybrid ring or rat-race circuit can be used as a 180° hybrid. For a 3-dB hybrid, the signal input at port 1 is split into ports 2 and 3 equally but 180° out of phase. Ports 1 and 4 are isolated. Similarly, ports 2 and 3 are isolated. The input signal at port 4 is split into ports 2 and 3 equally, but in phase. The characteristic impedance of the ring $Z_R = \sqrt{2}Z_0$ for a 3-dB hybrid [2, 3, 5], where Z_0 is the characteristic impedance of the input and output ports. A waveguide version of the hybrid ring called a magic-T is shown in Fig. 4.6.

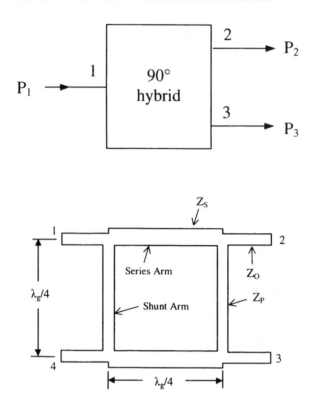

FIGURE 4.4   A 90° hybrid coupler. For a 3-dB hybrid, $Z_s = Z_0/\sqrt{2}$ and $Z_p = Z_0$.

A Wilkinson coupler is a two-way power divider or combiner. It offers broadband and equal-phase characteristics at each of its output ports. Figure 4.7 shows the one-section Wilkinson coupler, which consists of two quarter-wavelength sections. For a 3-dB coupler, the input at port 1 is split equally into two signals at ports 2 and 3. Ports 2 and 3 are isolated. A resistor of $2Z_0$ is connected between the two output ports to ensure the isolation [2, 3, 5]. For broadband operation, a multisection can be used. Unequal power splitting can be accomplished by designing different characteristic impedances for the quarter-wavelength sections and the resistor values [5]. The couplers can be cascaded to increase the number of output ports. Figure 4.8 shows a three-level one-to-eight power divider. Figure 4.9 shows the typical performance of a microstrip 3-dB Wilkinson coupler. Over the bandwidth of 1.8–2.25 GHz, the couplings at ports 2 and 3 are about 3.4 dB ($S_{21} \approx S_{31} \approx -3.4$ dB in Fig. 4.9). For the lossless case, $S_{21} = S_{31} = -3$ dB. Therefore, the insertion loss is about 0.4 dB. The isolation between ports 2 and 3 is over 20 dB.

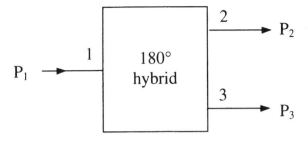

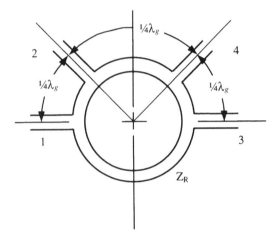

FIGURE 4.5 An $180°$ hybrid coupler. For a 3-dB hybrid, $Z_R = \sqrt{2}Z_0$.

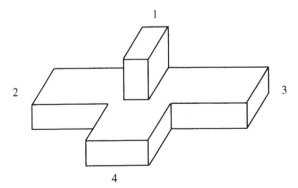

FIGURE 4.6 Waveguide magic-T circuit.

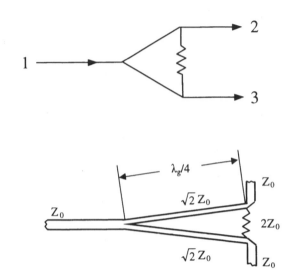

FIGURE 4.7   A 3-dB Wilkinson coupler.

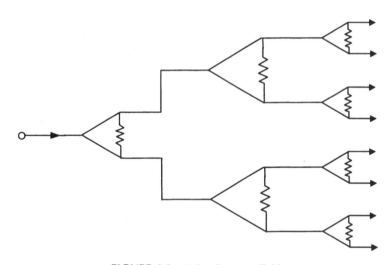

FIGURE 4.8   A 1 × 8 power divider.

4.3   RESONATORS, FILTERS, AND MULTIPLEXERS

Resonators and cavities are important components since they typically form filter networks. They are also used in controlling or stabilizing the frequency for oscillators, wave meters for frequency measurements, frequency discriminators, antennas, and measurement systems.

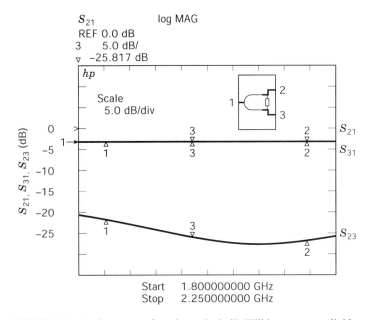

FIGURE 4.9 Performance of a microstrip 3-dB Wilkinson power divider.

Combinations of L and C elements form resonators. Figure 4.10 shows four types of combinations, and their equivalent circuits at the resonant frequencies are given in Fig. 4.11. At resonance, $Z = 0$, equivalent to a short circuit, and $Y' = 0$, equivalent to an open circuit. The resonant frequency is given by

$$\omega_0^2 = \frac{1}{LC} \tag{4.4}$$

or

$$f = f_r = \frac{1}{2\pi\sqrt{LC}} \tag{4.5}$$

In reality, there are losses (R and G elements) associated with the resonators. Figures 4.10a and c are redrawn to include these losses, as shown in Fig. 4.12. A quality factor Q is used to specify the frequency selectivity and energy loss. The unloaded Q is defined as

$$Q_0 = \frac{\omega_0(\text{time-averaged energy stored})}{\text{energy loss per second}} \tag{4.6a}$$

For a parallel resonator, we have

$$Q_0 = \frac{\omega_0(1/2)VV^*C}{(1/2)GVV^*} = \frac{\omega_0 C}{G} = \frac{R}{\omega_0 L} \tag{4.6b}$$

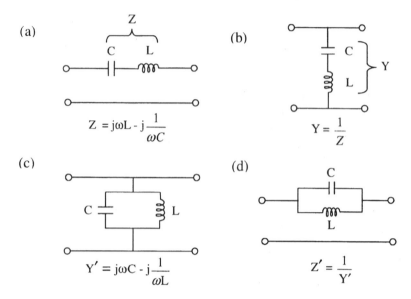

FIGURE 4.10 Four different basic resonant circuits.

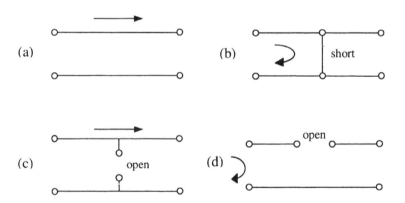

FIGURE 4.11 Equivalent circuits at resonance for the four resonant circuits shown in Fig. 4.10.

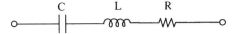

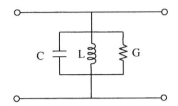

FIGURE 4.12 Resonators with lossy elements R and G.

For a series resonator, we have

$$Q_0 = \frac{\omega_0(1/2)II^*L}{(1/2)RII^*} = \frac{\omega_0 L}{R} = \frac{1}{\omega_0 CR} \tag{4.6c}$$

In circuit applications, the resonator is always coupled to the external circuit load. The loading effect will change the net resistance and consequently the quality factor [5]. A loaded Q is defined as

$$\frac{1}{Q_L} = \frac{1}{Q_0} + \frac{1}{Q_{ext}} \tag{4.7}$$

where Q_{ext} is the external quality factor due to the effects of external coupling. The loaded Q can be measured from the resonator frequency response [6]. Figure 4.13 shows a typical resonance response. The loaded Q of the resonator is

$$Q_L = \frac{f_0}{f_1 - f_2} \tag{4.8}$$

where f_0 is the resonant frequency and $f_1 - f_2$ is the 3-dB (half-power) bandwidth. The unloaded Q can be found from the loaded Q and the insertion loss IL (in decibels) at the resonance by the following equation [6]:

$$Q_0 = \frac{Q_L}{1 - 10^{-IL/20}} \tag{4.9}$$

The higher the Q value, the narrower the resonance response and the lower the circuit loss. A typical Q value for a microstrip resonator is less than 200, for a waveguide cavity is several thousand, for a dielectric resonator is around 1000, and for a crystal is over 5000. A superconductor can be used to lower the metallic loss and to increase the Q.

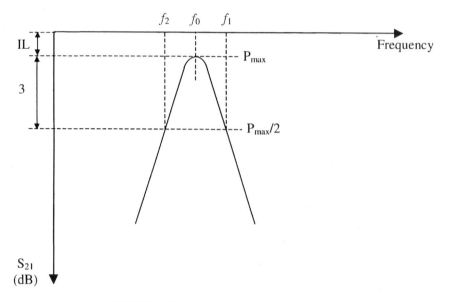

FIGURE 4.13   Resonator frequency response.

Commonly used resonators for microstrip circuits are open-end resonators, stub resonators, dielectric resonators, and ring resonators, as shown in Fig. 4.14. The boundary conditions force the circuits to have resonances at certain frequencies. For example, in the open-end resonator and stub resonator shown in Fig. 4.14, the voltage wave is maximum at the open edges. Therefore, the resonances occur for the open-end resonator when

$$l = n(\tfrac{1}{2}\lambda_g) \qquad n = 1, 2, 3, \ldots \tag{4.10}$$

For the open stub, the resonances occur when

$$l = n(\tfrac{1}{4}\lambda_g) \qquad n = 1, 2, 3, \ldots \tag{4.11}$$

For the ring circuit, resonances occur when

$$2\pi r = n\lambda_g \qquad n = 1, 2, 3, \ldots \tag{4.12}$$

The voltage (or E-field) for the first resonant mode ($n = 1$) for these circuits is shown in Fig. 4.15. From Eqs. (4.10)–(4.12), one can find the resonant frequencies by using the relation

$$\lambda_g = \frac{\lambda_0}{\sqrt{\varepsilon_{\text{eff}}}} = \frac{c}{f\sqrt{\varepsilon_{\text{eff}}}} \tag{4.13}$$

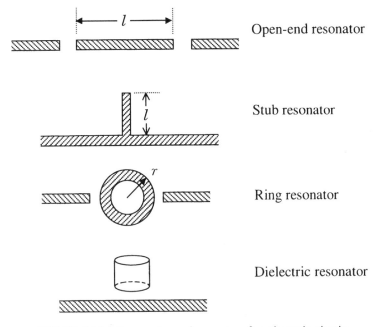

Open-end resonator

Stub resonator

Ring resonator

Dielectric resonator

FIGURE 4.14 Commonly used resonators for microstrip circuits.

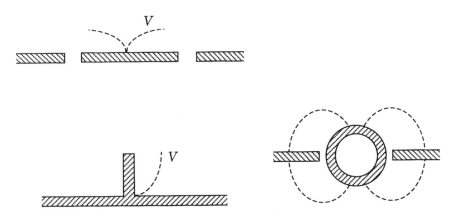

FIGURE 4.15 Voltage distribution for the first resonator mode.

Figure 4.16 shows the typical results for a loosely coupled ring resonator. Three resonances are shown for $n = 1, 2, 3$. The insertion loss is high because of the loose coupling [6].

One major application of the resonators is to build filters. There are four types of filters: the low-pass filter (LPF), bandpass filter (BPF), high-pass filter (HPF), and bandstop filter (BSF). Their frequency responses are shown in Fig. 4.17 [5]. An ideal

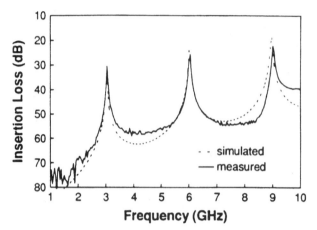

FIGURE 4.16 Microstrip ring resonator and its resonances.

filter would have perfect impedance matching, zero insertion loss in the passbands, and infinite rejection (attenuation or insertion loss) everywhere else. In reality, there is insertion loss in the passbands and finite rejection everywhere else. The two most common design characteristics for the passband are the maximum flat (Butterworth) response and equal-ripple (Chebyshev) response, as shown in Fig. 4.18, where A is the maximum attenuation allowed in the passband.

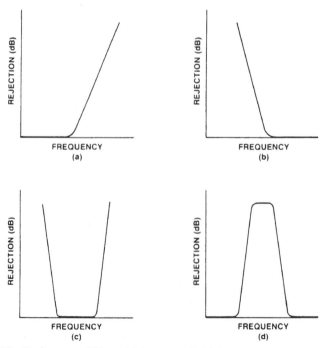

FIGURE 4.17 Basic types of filters: (*a*) low pass; (*b*) high pass; (*c*) bandpass; (*d*) bandstop.

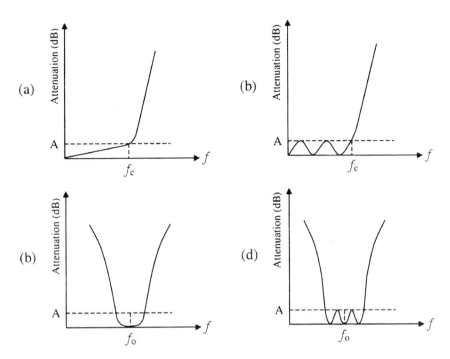

FIGURE 4.18   Filter response: (*a*) maximally flat LPF; (*b*) Chebyshev LPF; (*c*) maximally flat BPF; (*d*) Chebyshev BPF.

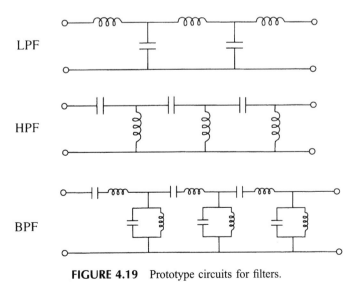

FIGURE 4.19   Prototype circuits for filters.

The prototype circuits for filters are shown in Fig. 4.19. In low frequencies, these circuits can be realized using lumped L and C elements. In microwave frequencies, different types of resonators and cavities are used to achieve the filter characteristics. Figure 4.20 shows some commonly used microstrip filter structures. The step impedance filter has low-pass characteristics; all others have bandpass characteristics. Figure 4.21 shows a parallel-coupled microstrip filter and its performance. The insertion loss (IL) in the passband around 5 GHz is about 2 dB, and the return loss (RL) is greater than 20 dB. The rejection at 4 GHz is over 20 dB and at 3 GHz is over 35 dB. The simulation can be done using a commercially available circuit simulator

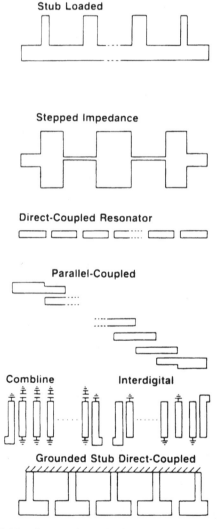

FIGURE 4.20   Commonly used microstrip filter structures [5].

or an electromagnetic simulator. For very narrow passband filters, surface acoustic wave (SAW) devices and dielectric resonators can be used.

The filter can be made electronically tunable by incorporating varactors into the filter circuits [1]. In this case, the passband frequency is tuned by varying the varactor bias voltages and thus the varactor capacitances. Active filters can be built by using active devices such as MESFETs in microwave frequencies and CMOS in RF. The active devices provide negative resistance and compensate for the losses of the filters. Active filters could have gains instead of losses.

A frequency multiplexer is a component that separates or combines signals in different frequency bands (Fig. 4.22a). It is used in frequency division multiple

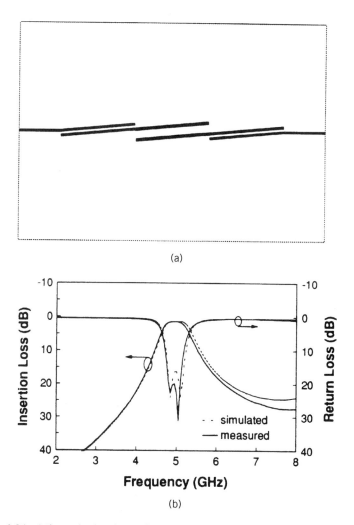

FIGURE 4.21   Microstrip bandpass filter and its performance: (*a*) circuit layout; (*b*) simulated and measured results.

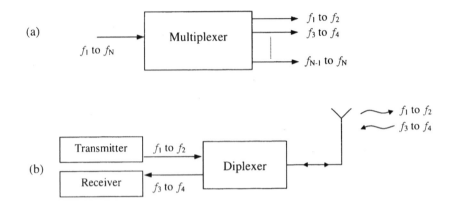

FIGURE 4.22   Multiplexer and diplexer: (*a*) a multiplexer is used to separate many different frequency bands: (*b*) a diplexer is used to separate the transmitting and receiving signals in a communication system.

access (FDMA) to divide a frequency band into many channels or users in a communication system. Guard bands are normally required between the adjacent channels to prevent interference. A filter bank that consists of many filters in parallel can be used to accomplish the frequency separation. A diplexer is a component used to separate two frequency bands. It is commonly used as a duplexer in a transceiver (transmitter and receiver) to separate the transmitting and receiving frequency bands. Figure 4.22*b* shows a diplexer used for this function.

4.4   ISOLATORS AND CIRCULATORS

Isolators and circulators are nonreciprocal devices. In many cases, they are made with ferrite materials. The nonreciprocal electrical properties cause that the transmission coefficients passing through the device are not the same for different directions of propagation [2]. In an isolator, almost unattenuated transmission from port 1 to port 2 is allowed, but very high attenuation exists in the reverse direction from port 2 to port 1, as shown in Fig. 4.23. The isolator is often used to couple a microwave signal source (oscillator) to the external load. It allows the available power to be delivered to the load but prevents the reflection from the load transmitted back to the source. Consequently, the source always sees a matched load, and the effects of the load on the source (such as frequency pulling or output power variation) are minimized. A practical isolator will introduce an insertion loss for the power transmitted from port 1 to port 2 and a big but finite isolation (rejection) for the power transmitted from port 2 to port 1. Isolation can be increased by cascading two isolators in series. However, the insertion loss is also increased.

Example 4.2   The isolator shown in Fig. 4.23*a* has an insertion loss α_L of 1 dB and an isolation α_I of 30 dB over the operation bandwidth. (a) What is the output power

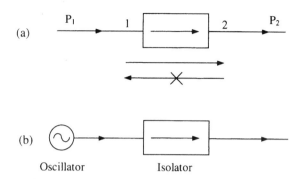

(a)

(b)   Oscillator   Isolator

FIGURE 4.23   Isolator and its applications: (*a*) isolator allows power to flow in one direction only; (*b*) isolator is used to protect an oscillator.

P_2 at port 2 if the input power at port 1 is $P_1 = 10$ mW? (b) What is the output power P_1 at port 1 if the input power at port 2 is $P_2 = 10$ mW?

Solution

(a)
$$P_2 = P_1 - \alpha_L = 10 \text{ dBm} - 1 \text{ dB} = 9 \text{ dBm}$$
$$= 7.94 \text{ mW}$$

(b)
$$P_1 = P_2 - \alpha_I = 10 \text{ dBm} - 30 \text{ dB} = -20 \text{ dBm}$$
$$= 0.01 \text{ mW}$$ ∎

A circulator is a multiport device for signal routing. Figure 4.24 shows a three-port circulator. A signal incident in port 1 is coupled into port 2 only, a signal incident in port 2 is coupled into port 3 only, and a signal incident in port 3 is coupled into port 1 only. The signal traveling in the reverse direction is the leakage determined by the isolation of the circulator. A circulator is a useful component for signal routing or separation, and some applications are shown in Fig. 4.25. A terminated circulator can be used as an isolator (Fig. 4.25*a*). The reflection from port 2 is dissipated in the termination at port 3 and will not be coupled into port 1. Figure 4.25*b* shows that a circulator can be used as a duplexer in a transceiver to separate the transmitted and received signals. The transmitted and received signals can have the same or different frequencies. This arrangement is quite popular for radar applications. The circuit shown in Fig. 4.25*c* is a fixed or a variable phase shifter. By adjusting the length *l* of a transmission line in port 2, one can introduce a phase shift of $2\beta l$ between ports 1 and 3. The length *l* can be adjusted by using a sliding (tunable) short. A circulator can be used to build an injection locked or a stable amplifier using a two-terminal solid-state active device such as an IMPATT diode or a Gunn device [1]. The circulator is used to separate the input and output ports in this case, as shown in Fig. 4.25*d*.

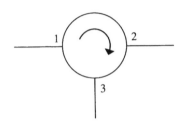

FIGURE 4.24 Three-port circulator.

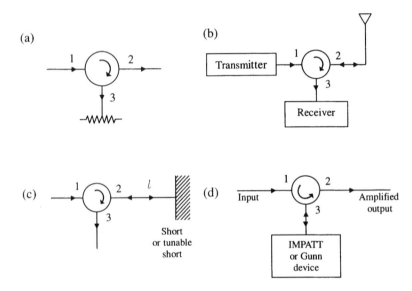

FIGURE 4.25 Some applications of circulators: (*a*) as an isolator; (*b*) as a duplexer; (*c*) as a phase shifter; (*d*) as an amplifier circuit.

4.5 DETECTORS AND MIXERS

A detector is a device that converts an RF/microwave signal into a DC voltage or that demodulates a modulated RF/microwave signal to recover a modulating low-frequency information-bearing signal. Detection is accomplished by using a nonlinear I–V device. A p–n junction or a Schottky-barrier junction (metal–semiconductor junction) has a nonlinear I–V characteristic, as shown in Fig. 4.26. The characteristic can be given by [1]

$$i = a_1 v + a_2 v^2 + a_3 v^3 + \cdots \tag{4.14}$$

If a continuous wave is incident to the detector diode, as shown in Fig. 4.27a, we have

$$v = A \cos \omega_{RF} t \quad \text{or} \quad v = A \sin \omega_{RF} t \tag{4.15}$$

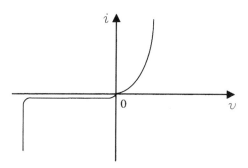

FIGURE 4.26 Nonlinear I–V characteristics.

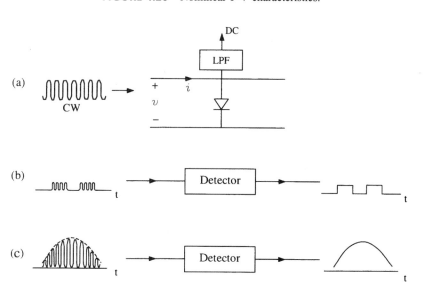

FIGURE 4.27 Detectors are used to (*a*) convert a CW RF signal to DC output, (*b*) demondulate a pulse-modulated RF carrier, and (*c*) demodulate an analog-modulated RF carrier.

The first two terms will give

$$i = a_1 A \cos \omega_{RF} t + a_2 A^2 \cos^2 \omega_{RF} t$$
$$= a_1 A \cos \omega_{RF} t + \tfrac{1}{2} a_2 A^2 + \tfrac{1}{2} a_2 A^2 \cos 2\omega_{RF} t \qquad (4.16)$$

A DC current appears at the output of a low-pass filter:

$$i_{DC} = \tfrac{1}{2} a_2 A^2 \propto A^2 \qquad (4.17)$$

The detector is normally operating in the square-law region with the DC current proportional to the square of the incident RF signal [7]. If the incident RF signal is pulse modulated, as shown in Fig. 4.27b, then DC currents appear only when there are carrier waves. The output is the demodulated signal (modulating signal) of the pulse-modulated carrier signal. Similarly, the detector's output for an analog-modulated signal is the modulating low-frequency signal bearing the analog information.

The performance of a detector is judged by its high sensitivity, good VSWR, high dynamic range, low loss, and wide operating bandwidth. The current sensitivity of a detector is defined as

$$\beta_i = \frac{i_{DC}}{P_{in}} \tag{4.18}$$

where P_{in} is the incident RF power and i_{DC} is the detector output DC current.

Since the baseband modulating signal usually contains frequencies of less than 1 MHz, the detector suffers from $1/f$ noise (flicker noise). The sensitivity of the RF/microwave receiver can be greatly improved by using the heterodyne principle to avoid the $1/f$ noise. In heterodyne systems, the initial baseband frequency is converted up to a higher transmitted carrier frequency, and then the process is reversed at the receiver. The frequency conversions are done by mixers (upconverters and downconverters). In the downconverter, as shown in Fig. 4.28, the high-frequency received signal (RF) is mixed with a local oscillator (LO) signal to generate a difference signal, which is called the intermediate-frequency (IF) signal. The IF signal can be amplified and detected/demodulated. It can also be further downconverted to a lower frequency IF before detection or demodulation. The upconverter is used to generate a high-frequency RF signal for transmission from a low-frequency information-bearing IF signal. The upconverter is used in a transmitter and the downconverter in a receiver.

The input voltage to the downconverter is given by

$$v = A \sin \omega_{RF}t + B \sin \omega_{LO}t \tag{4.19}$$

Substituting this into Eq. (4.14) gives

$$
\begin{aligned}
i = {}& a_1(A \sin \omega_{RF}t + B \sin \omega_{LO}t) \\
& + a_2(A^2 \sin^2 \omega_{RF}t + 2AB \sin \omega_{RF}t \sin \omega_{LO}t + B^2 \sin^2 \omega_{LO}t) \\
& + a_3(A^3 \sin^3 \omega_{RF}t + 3A^2B \sin^2 \omega_{RF}t \sin \omega_{LO}t \\
& + 3AB^2 \sin \omega_{RF}t \sin^2 \omega_{LO}t + B^3 \sin^3 \omega_{LO}t) + \cdots
\end{aligned} \tag{4.20}
$$

Because the term $2AB \sin \omega_{RF}t \sin \omega_{LO}t$ is just the multiplication of the two input signals, the mixer is often referred as a multiplier for two signals, as shown in Fig. 4.29.

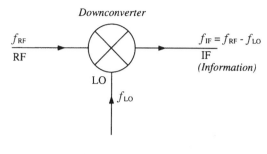

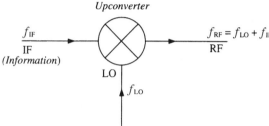

FIGURE 4.28   Downconverter and upconverter.

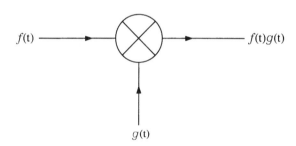

FIGURE 4.29   Multiplication of two input signals by a mixer.

Using the trigonometric identities, the following frequency components result from (4.20):

$$a_1 v \rightarrow \omega_{RF}, \omega_{LO}$$
$$a_2 v^2 \rightarrow 2\omega_{RF}, \omega_{RF} \pm \omega_{LO}, 2\omega_{LO}$$
$$a_3 v^3 \rightarrow 3\omega_{RF}, 2\omega_{RF} \pm \omega_{LO}, 2\omega_{LO} \pm \omega_{RF}, 3\omega_{LO}, \omega_{RF}, \omega_{LO}$$
$$\vdots$$

For the downconverter, a low-pass filter is used in the mixer to extract the IF signal ($\omega_{RF} - \omega_{LO}$ or $\omega_{LO} - \omega_{RF}$). All other frequency components are trapped and eventually converted to the IF signal or dissipated as heat. For the upconverter, a bandpass filter is used to pass $\omega_{IF} + \omega_{LO}$.

The conversion loss for a downconverter is defined as

$$L_c \text{ (in dB)} = 10 \log \frac{P_{RF}}{P_{IF}} \tag{4.21}$$

where P_{RF} is the input RF signal power to the mixer and P_{IF} is the output IF signal power.

A good mixer requires low conversion loss, a low noise figure, low VSWRs for the RF, IF, and LO ports, good isolation between any two of the RF, IF, and LO ports, good dynamic range, a high 1-dB compression point, a high third-order intercept point, and low intermodulation. Definitions of dynamic range, third-order intercept point, 1-dB compression point, and intermodulation will be given in Chapter 5. As an example of mixer performance, a 4–40-GHz block downconverter from Miteq has the following typical specifications [8]:

RF frequency range	4–40 GHz
LO frequency range	4–42 GHz
IF frequency range	0.5–20 GHz
RF VSWR	2.5
IF VSWR	2.5
LO VSWR	2.0
LO-to-RF isolation	20 dB
LO-to-IF isolation	25 dB
RF-to-IF isolation	30 dB
Conversion loss	10 dB
Single-sideband noise figure (at 25°C)	10.5 dB
Input power at 1 dB compression	+5 dBm
Input power at third-order intercept point	+15 dBm
LO power requirement	+10 to +13 dBm

Note that the noise figure is approximately equal to the conversion loss for a lossy element (as described in Chapter 5). The mixer normally consists of one or more nonlinear devices and associated filtering circuits. The circuits can be realized by using a microstrip line or waveguide [1]. The same $p–n$ junction or Schottky-barrier junction diodes employed for detectors can be used for mixers. The use of transistors (e.g., MESFET, HEMT) as the nonlinear devices has the advantage of providing conversion gain instead of conversion loss.

4.6 SWITCHES, PHASE SHIFTERS, AND ATTENUATORS

Switches, phase shifters, and attenuators are control devices that provide electronic control of the phase and amplitude of RF/microwave signals. The control devices can be built by using ferrites or solid-state devices ($p–i–n$ diodes or FETs) [1, 7]. Phase shifting and switching with ferrites are usually accomplished by changing the magnetic permeability, which occurs with the application of a magnetic biasing field. Ferrite control devices are heavy, slow, and expensive. Solid-state control devices, on the other hand, are small, fast, and inexpensive. The ferrite devices do have some advantages such as higher power handling and lower loss. Table 4.1 gives the comparison between ferrite and $p-i-n$ diode control devices [1]. It should be mentioned that the use of FETs or transistors as control devices could provide gain instead of loss.

Switches are widely used in communication systems for time multiplexing, time division multiple access (see Chapter 10), pulse modulation, channel switch in the channelized receiver, transmit/receive (T/R) switch for a transceiver, and so on. Figure 4.30 illustrates these applications. A switch can be classified as single pole, single throw (SPST), single pole, double throw (SPDT), single pole, triple throw (SP3T), and so on, as shown in Fig. 4.31. Ideally, if the switch is turned on, all signal power will pass through without any attenuation. When the switch is off, all power will be rejected and no power leaks through. In reality, there is some insertion loss when the switch is on and some leakage when the switch is off. From Fig. 4.32 the insertion loss and the isolation are given by

When the switch is on,

$$\text{Insertion loss} = \alpha_L = 10 \log \frac{P_{\text{in}}}{P_{\text{out}}} \qquad (4.22)$$

When the switch is off,

$$\text{Isolation} = \alpha_I = 10 \log \frac{P_{\text{in}}}{P_{\text{out}}} \qquad (4.23)$$

A good switch should have low insertion loss and high isolation. Other desired features depending on applications are fast switching speed, low switching current, high power-handling capability, small size, and low cost. For a solid-state switch, the switching is accomplished by the two device impedance states obtained from two

TABLE 4.1   Comparison between Ferrite and $p-i-n$ Diode Control Devices

Parameter	Ferrite	$p-i-n$
Speed	Low (msec)	High (μsec)
Loss	Low (0.2 dB)	High (0.5 dB/diode)
Cost	High	Low
Weight	Heavy	Light
Driver	Complicated	Simple
Size	Large	Small
Power handling	High	Low

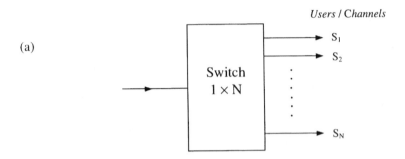

(a)

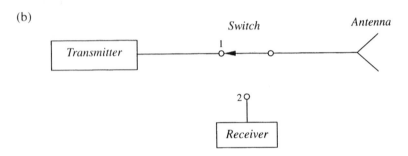

(b)

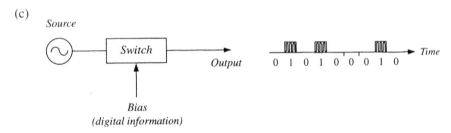

(c)

FIGURE 4.30 Applications of switches: (*a*) channel switch or time multiplexing; (*b*) T/R switch or duplexer; (*c*) pulse modulator.

different bias states [1]. For one state, the device acts as a short circuit, and for the other, as an open circuit.

One major application of switches is to build phase shifters. Figure 4.33 shows a switched-line phase shifter and its realization using p–i–n diodes [1]. When the bias is positive, the signal flows through the upper line with a path length l_1. If the bias is negative, the signal flows through the lower line with a path length l_2. The phase difference between the two bias states is called a differential phase shift, given by

$$\Delta\phi = \frac{2\pi}{\lambda_g}(l_1 - l_2) \tag{4.24}$$

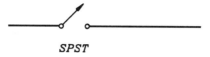

SPST

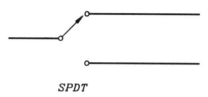

SPDT

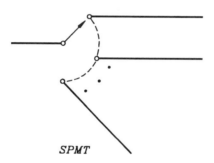

SPMT

FIGURE 4.31 Switches and their output ports.

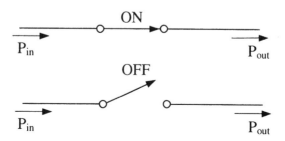

FIGURE 4.32 Switch in on and off positions.

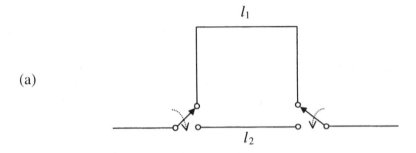

(a)

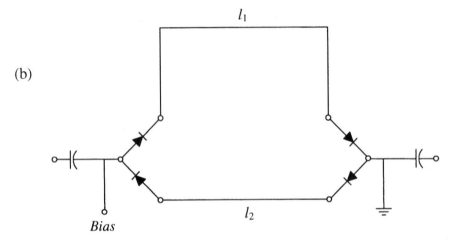

(b)

FIGURE 4.33   Switch-line phase shifter: (*a*) schematic diagram; (*b*) construction using *p–i–n* diodes.

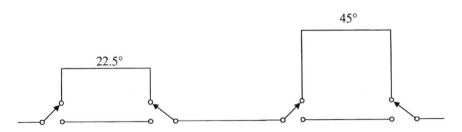

FIGURE 4.34   Two-bit phase shifter.

This phase shifter provides two phase states and is a 1-bit phase shifter. For more states, one can cascade two or more 1-bit phase shifters. Figure 4.34 illustrates an example of a 2-bit phase shifter. Four differential phase states result from switching the four SPDT switches. These phases are 0° (reference), 22.5°, 45°, and 67.5°. One major application of phase shifters is in phased-array antennas.

Instead of operating in two states, on and off as in a switch, one can vary the bias continuously. The device impedance is then varied continuously and the attenuation (insertion loss) is changed continuously. The component becomes a variable attenuator or electronically tunable attenuator. One application of the variable attenuator is automatic gain control used in many receiver systems.

4.7 OSCILLATORS AND AMPLIFIERS

Oscillators and amplifiers are active components. The component consists of a solid-state device (transistor, FET, IMPATT, Gunn, etc.) that generates a negative resistance when it is properly biased. A positive resistance dissipates RF power and introduces losses. In contrast, a negative resistance generates RF power from the DC bias supplied to the active solid-state device. Figure 4.35 shows a general oscillator circuit, where Z_D is the solid-state device impedance and Z_C is the circuit impedance looking at the device terminals (driving point) [1]. The impedance transformer network includes the device package and embedding circuit. The circuit impedance seen by the device is

$$Z_c(f) = R_c(f) + jX_c(f) \tag{4.25}$$

For the oscillation to occur, two conditions need to be satisfied,

$$\mathrm{Im}(Z_D) = -\mathrm{Im}(Z_C) \tag{4.26}$$

$$|\mathrm{Re}(Z_D)| \geq \mathrm{Re}(Z_C) \tag{4.27}$$

where Im and Re mean imaginary and real parts, respectively. The real part of Z_D is negative for a negative resistance. The circuit impedance is only a function of

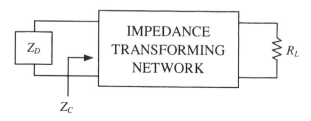

FIGURE 4.35   General oscillator circuit.

frequency. The device impedance is generally a function of frequency, bias current, RF current, and temperature. Thus at the oscillating frequency f_0, we have

$$R_C(f_0) \leq |R_D(f_0, I_0, I_{RF}, T)| \tag{4.28}$$

$$X_C(f_0) + X_D(f_0, I_0, I_{RF}, T) = 0 \tag{4.29}$$

Equation (4.28) states that the magnitude of the negative device resistance is greater than the circuit resistance. Therefore, there is a net negative resistance in the overall circuit. Equation (4.29) indicates that the oscillating frequency is the circuit resonant frequency since the total reactance (or admittance) equals zero at resonance. For a transistor or any three-terminal solid-state device, Z_D is replaced by the transistor and a termination, as shown in Fig. 4.36. The same oscillation conditions given by Eqs. (4.28) and (4.29) are required.

Oscillators are used as sources in transmitters and as local oscillators in upconverters and downconverters. System parameters of interest include power output, DC-to-RF efficiency, noise, stability, frequency tuning range, spurious signals, frequency pulling, and frequency pushing. These parameters will be discussed in detail in Chapter 6.

An amplifier is a component that provides power gain to the input signal to the amplifier. As shown in Fig. 4.37, P_{in} is the input power and P_{out} is the output power. The power gain is defined as

$$G = \frac{P_{out}}{P_{in}} \tag{4.30}$$

or

$$G \text{ (in dB)} = 10 \log \frac{P_{out}}{P_{in}} \tag{4.31}$$

Amplifiers can be cascaded to provide higher gain. For example, for two amplifiers with gain G_1 and G_2 in cascade, the total gain equals $G_1 G_2$. The amplifier used in

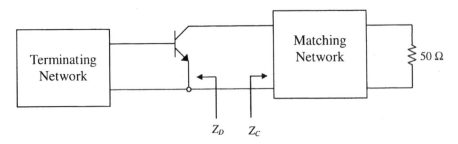

FIGURE 4.36 Transistor oscillator.

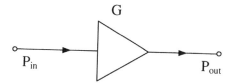

FIGURE 4.37 Amplifier with power gain G.

the last stage of a transmitter provides high power output and is generally called a power amplifier (PA). The amplifier used in the receiver normally has a low noise figure and is called a low-noise amplifier (LNA). An amplifier can be constructed by designing the input and output matching network to match an active solid-state device. Figure 4.38 shows a transistor amplifier circuit [1]. The important design considerations for an amplifier are gain, noise, bandwidth, stability, and bias arrangement. An amplifier should not oscillate in the operating bandwidth. The stability of an amplifier is its resistance to oscillation. An unconditionally stable amplifier will not oscillate under any passive termination of the input and output circuits.

For a power amplifier, desired system parameters are high power output, high 1-dB compression point, high third-order intercept point, large dynamic range, low intermodulation, and good linearity. Most of these parameters will be defined and discussed in Chapters 5 and 6. For battery operating systems, high power added efficiency (PAE) is also important. The PAE is defined as

$$PAE = \frac{P_{out} - P_{in}}{P_{DC}} \times 100\% \qquad (4.32)$$

where P_{DC} is the DC bias power. Power added efficiencies of over 50% are routinely achievable for transistor amplifiers.

Table 4.2 gives the typical performance of a Miteq amplifier [8].

Example 4.3 In the system shown in Fig. 4.39, calculate the output power in milliwatts when (a) the switch is on and (b) the switch is off. The switch has an insertion loss of 1 dB and an isolation of 30 dB.

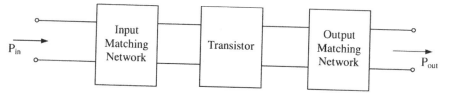

FIGURE 4.38 Transistor amplifier circuit.

TABLE 4.2 Performance of Miteq Amplifier Model MPN2-01000200-28P

Operating frequency	1–2 GHz
Gain	27 dB minimum
Gain flatness	±1.5 dB maximum
Noise figure	1.5 dB maximum
VSWR	2.0 maximum
Output 1 dB compression point	+28 dBm
Output third-order intercept point	+40 dBm

Solution

$$P_{in} = 0.001 \text{ mW} = -30 \text{ dBm}$$

For the switch, $\alpha_L = 1$ dB, $\alpha_I = 30$ dB:

(a) When the switch is ON, we have

$$P_{out} = P_{in} - L - L_c - \alpha_L + G_1 + G_2$$
$$= -30 \text{ dBm} - 1 \text{ dB} - 4 \text{ dB} - 1 \text{ dB} + 10 \text{ dB} + 30 \text{ dB}$$
$$= +4 \text{ dBm} = 2.51 \text{ mW}$$

(b) When the switch is OFF, we have

$$P_{out} = P_{in} - L - L_c - \alpha_1 + G_1 + G_2$$
$$= -30 \text{ dBm} - 1 \text{ dB} - 4 \text{ dB} - 30 \text{ dB} + 10 \text{ dB} + 30 \text{ dB}$$
$$= -25 \text{ dBm} = 0.00316 \text{ mW} \qquad \blacksquare$$

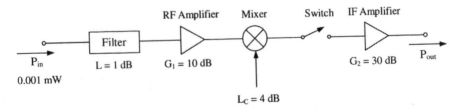

FIGURE 4.39 Receiver system.

4.8 FREQUENCY MULTIPLIERS AND DIVIDERS

A frequency multiplier is used to generate the output signal with a frequency that is a multiple of the input signal frequency, as shown in Fig. 4.40. If the input frequency is f_0, the output frequency is nf_0, and n could be any of 2, 3, 4, ... When $n = 2$, it is a $\times 2$ multiplier, or a doubler. When $n = 3$, it is a $\times 3$ multiplier, or a tripler. The multiplier consists of a low-pass filter, a nonlinear device such as a step recovery diode or a varactor, and input- and output-matching networks. Figure 4.41 shows a block diagram [1]. The low-pass filter, located in the input side, passes the fundamental signal and rejects all higher harmonics. The varactor is the nonlinear device that produces harmonics. The bandpass or high-pass filter at the output side

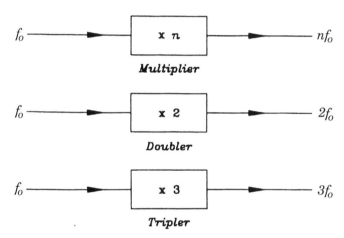

FIGURE 4.40 Frequency multipliers.

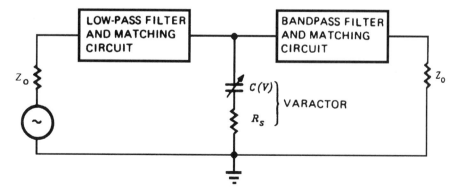

FIGURE 4.41 Multiplier circuit schematic: Z_0 = load impedance or characteristic impedance of transmission line; $C(V)$, R_s = variable capacitance and series resistance of varactor.

passes only the desired harmonic and rejects all other signals. The conversion efficiency (η) and conversion loss (L_c) are defined as

$$\eta = \frac{P_{out}}{P_{in}} \times 100\% \qquad (4.33)$$

$$L_c \text{ (in dB)} = 10 \log \frac{P_{in}}{P_{out}} \qquad (4.34)$$

where P_{in} is the input power of the fundamental frequency and P_{out} is the output power of the desired harmonic. Frequency multipliers have been built up to millimeter-wave and submillimeter-wave frequencies [7].

Frequency dividers are commonly used in phase-locked loops (PLLs) and frequency synthesizers (Chapter 6). A frequency divider generates a signal with a frequency that is $1/N$ of the input signal frequency, where $N = 2, 3, 4, \ldots$. Figure 4.42 shows a symbol of a frequency divider and its input and output frequencies. Frequency division may be achieved in many ways. One example is to use the mixer-with-feedback method shown in Fig. 4.43. This is also called a regenerative divider. The mixer output frequency is

$$f_0 - f_0\left(\frac{N-1}{N}\right) = \frac{f_0}{N} \qquad (4.35)$$

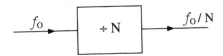

FIGURE 4.42   Frequency divider.

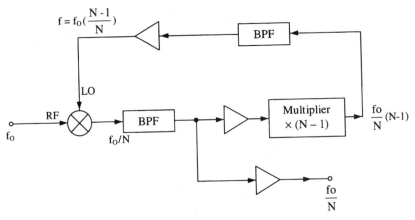

FIGURE 4.43   Regenerative frequency divider.

The maximum division ratio depends on the selectivity of the bandpass filter following the mixer. The amplifiers used in Fig. 4.43 are to boost the signal levels.

PROBLEMS

4.1   A 6-dB microstrip directional coupler is shown in Fig. P4.1. The coupling is 6 dB, and the directivity is 30 dB. If the input power is 10 mW, calculate the output power at ports 2, 3, and 4. Assume that the coupler is lossless.

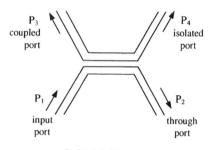

P_3
coupled
port

P_4
isolated
port

P_1
input
port

P_2
through
port

FIGURE P4.1

4.2   A three-way power divider has an insertion loss of 0.5 dB. If the input power is 0 dBm, what is the output power in dBm and milliwatts at any one of the output ports?

4.3   A bandpass filter (maximum flat characteristics) has the following specifications:

Passband from 9 to 10 GHz

Insertion loss 0.5 dB maximum

Off-band rejection:

At 8 GHz	IL = 30 dB
At 8.5 GHz	IL = 20 dB
At 10.5 GHz	IL = 15 dB
At 11 GHz	IL = 25 dB

(a) Plot the characteristics in log-scale (i.e., decibels vs. frequency).

(b) Plot the characteristics in regular scale (i.e., magnitude vs. frequency).

4.4   A downconverter has a conversion loss of 4.17 dB and RF and LO isolation of 20 dB. If the RF input power is 0 dBm, what are the IF output power and the RF power leaked into the LO port?

4.5   A switch has an insertion loss of 0.4 dB and isolation of 25 dB. If the input power is 1 mW, what are the output power levels when the switch is on and off?

4.6   In the system shown in Fig. P4.6, assume that all components are matched to the transmission lines. Calculate the power levels P_A, P_B, and P_C in milliwatts (P_A, P_B, and P_C are shown in the figure).

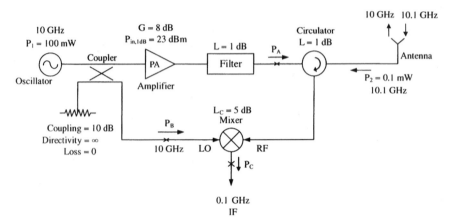

FIGURE P4.6

4.7   A transmitter is connected through a switch and a cable to an antenna (Fig. P4.7). If the switch has an insertion loss of 1 dB and an isolation of 20 dB, calculate the power radiated when the switch is in the on and off positions. The cable has an insertion loss of 2 dB. The antenna has an input VSWR of 2 and 90% radiating efficiency.

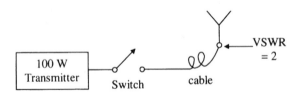

FIGURE P4.7

4.8   A transceiver is shown in Fig. P4.8. In the transmitting mode, the transmitter transmits a signal of 1 W at 10 GHz. In the receiving mode, the antenna receives a signal P_2 of 1 mW at 12 GHz. A switch is used as a duplexer. The switch has an insertion loss of 2 dB and isolation of 40 dB. The bandpass filter (BPF) has an insertion loss of 2 dB at 12 GHz and a rejection of 30 dB at 10 GHz. Calculate (a) the transmitted power P_1 at the antenna input, (b) the 10-GHz leakage power at the receiver input, (c) the 12-GHz received power at the receiver input.

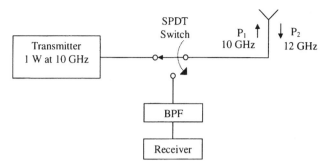

FIGURE P4.8

4.9 A system is shown in Fig. P4.9. The switch is connected to position 1 when the system is transmitting and to position 2 when the system is receiving. Calculate (a) the transmitting power P_t in milliwats and (b) the receiver output power P_{out} in milliwatts. Note that $P_r = 0.001$ mW, and the switch has an insertion loss of 1 dB.

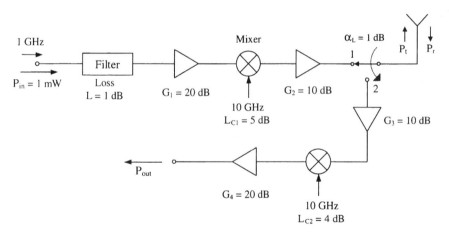

FIGURE P4.9

4.10 In the system shown in Fig. P4.10, calculate the power levels P_A, P_B, P_C, and P_D in milliwatts. Assume that all components are matched to the transmission lines.

4.11 List all states of phase shifts available in the circuit shown in Fig. P4.11.

4.12 Redraw Fig. 4.43 for a regenerative frequency divider for $N = 2$. What is the output power for this divider if the input power is 1 mW? Assume that the mixer conversion loss is 10 dB, the bandpass filter insertion loss is 4 dB, and the amplifier gain is 18 dB.

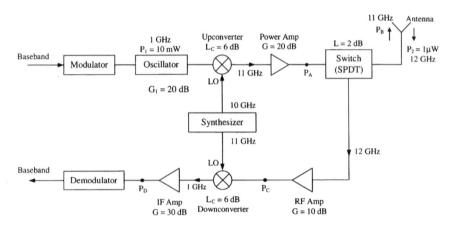

FIGURE P4.10

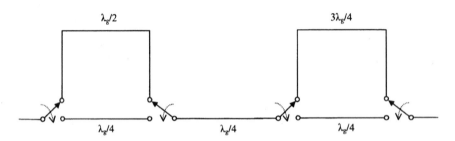

FIGURE P4.11

REFERENCES

1. K. Chang, *Microwave Solid-State Circuits and Applications*, John Wiley & Sons, New York, 1994.

2. R. E. Collin, *Foundation for Microwave Engineering*, 2nd ed., McGraw-Hill, New York, 1992.

3. D. M. Pozar, *Microwave Engineering*, 2nd ed., John Wiley & Sons, New York, 1998.

4. S. Y. Liao, *Microwave Devices and Circuits*, 3rd ed., Prentice-Hall, Englewood Cliffs, NJ, 1990.

5. K. Chang, Ed., *Handbook of Microwave and Optical Components*, Vol. 1, *Microwave Passive and Antenna Components*, John Wiley & Sons, New York, 1989.

6. K. Chang, *Microwave Ring Circuits and Antennas*, John Wiley & Sons, New York, 1996, Ch. 6.

7. K. Chang, Ed., *Handbook of Microwave and Optical Components*, Vol. 2, *Microwave Solid-State Components*, John Wiley & Sons, New York, 1990.

8. *Miteq Catalog*, Miteq, Hauppauge, NY.

CHAPTER FIVE

Receiver System Parameters

5.1 TYPICAL RECEIVERS

A receiver picks up the modulated carrier signal from its antenna. The carrier signal is downconverted, and the modulating signal (information) is recovered. Figure 5.1 shows a diagram of typical radio receivers using a double-conversion scheme. The receiver consists of a monopole antenna, an RF amplifier, a synthesizer for LO signals, an audio amplifier, and various mixers, IF amplifiers, and filters. The input signal to the receiver is in the frequency range of 20–470 MHz; the output signal is an audio signal from 0 to 8 kHz. A detector and a variable attenuator are used for automatic gain control (AGC). The received signal is first downconverted to the first IF frequency of 515 MHz. After amplification, the first IF frequency is further downconverted to 10.7 MHz, which is the second IF frequency. The frequency synthesizer generates a tunable and stable LO signal in the frequency range of 535–985 MHz to the first mixer. It also provides the LO signal of 525.7 MHz to the second mixer.

Other receiver examples are shown in Fig. 5.2. Figure 5.2a shows a simplified transceiver block diagram for wireless communications. A T/R switch is used to separate the transmitting and receiving signals. A synthesizer is employed as the LO to the upconverter and downconverter. Figure 5.2b is a mobile phone transceiver (transmitter and receiver) [1]. The transceiver consists of a transmitter and a receiver separated by a filter diplexer (duplexer). The receiver has a low noise RF amplifier, a mixer, an IF amplifier after the mixer, bandpass filters before and after the mixer, and a demodulator. A frequency synthesizer is used to generate the LO signal to the mixer.

Most components shown in Figs. 5.1 and 5.2 have been described in Chapters 3 and 4. This chapter will discuss the system parameters of the receiver.

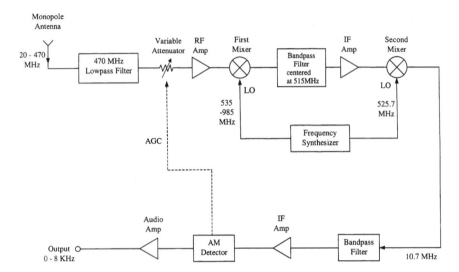

FIGURE 5.1 Typical radio receiver.

5.2 SYSTEM CONSIDERATIONS

The receiver is used to process the incoming signal into useful information, adding minimal distortion. The performance of the receiver depends on the system design, circuit design, and working environment. The acceptable level of distortion or noise varies with the application. Noise and interference, which are unwanted signals that appear at the output of a radio system, set a lower limit on the usable signal level at the output. For the output signal to be useful, the signal power must be larger than the noise power by an amount specified by the required minimum signal-to-noise ratio. The minimum signal-to-noise ratio depends on the application, for example, 30 dB for a telephone line, 40 dB for a TV system, and 60 dB for a good music system.

To facilitate the discussion, a dual-conversion system as shown in Fig. 5.3 is used. A preselector filter (Filter 1) limits the bandwidth of the input spectrum to minimize the intermodulation and spurious responses and to suppress LO energy emission. The RF amplifier will have a low noise figure, high gain, and a high intercept point, set for receiver performance. Filter 2 is used to reject harmonics generated by the RF amplifier and to reject the image signal generated by the first mixer. The first mixer generates the first IF signal, which will be amplified by an IF amplifier. The IF amplifier should have high gain and a high intercept point. The first LO source should have low phase noise and sufficient power to pump the mixer. The receiver system considerations are listed below.

1. *Sensitivity.* Receiver sensitivity quantifies the ability to respond to a weak signal. The requirement is the specified signal-noise ratio (SNR) for an analog receiver and bit error rate (BER) for a digital receiver.

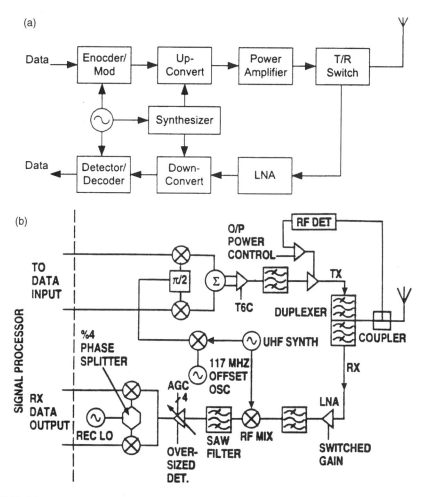

FIGURE 5.2 (a) Simplified transceiver block diagram for wireless communications. (b) Typical mobile phone transceiver system. (From reference [1], with permission from IEEE.)

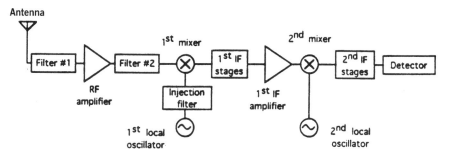

FIGURE 5.3 Typical dual-conversion receiver.

2. *Selectivity.* Receiver selectivity is the ability to reject unwanted signals on adjacent channel frequencies. This specification, ranging from 70 to 90 dB, is difficult to achieve. Most systems do not allow for simultaneously active adjacent channels in the same cable system or the same geographical area.

3. *Spurious Response Rejection.* The ability to reject undesirable channel responses is important in reducing interference. This can be accomplished by properly choosing the IF and using various filters. Rejection of 70 to 100 dB is possible.

4. *Intermodulation Rejection.* The receiver has the tendency to generate its own on-channel interference from one or more RF signals. These interference signals are called intermodulation (IM) products. Greater than 70 dB rejection is normally desirable.

5. *Frequency Stability.* The stability of the LO source is important for low FM and phase noise. Stabilized sources using dielectric resonators, phase-locked techniques, or synthesizers are commonly used.

6. *Radiation Emission.* The LO signal could leak through the mixer to the antenna and radiate into free space. This radiation causes interference and needs to be less than a certain level specified by the FCC.

5.3 NATURAL SOURCES OF RECEIVER NOISE

The receiver encounters two types of noise: the noise picked up by the antenna and the noise generated by the receiver. The noise picked up by the antenna includes sky noise, earth noise, atmospheric (or static) noise, galactic noise, and man-made noise. The sky noise has a magnitude that varies with frequency and the direction to which the antenna is pointed. Sky noise is normally expressed in terms of the noise temperature (T_A) of the antenna. For an antenna pointing to the earth or to the horizon $T_A \simeq 290$ K. For an antenna pointing to the sky, its noise temperature could be a few kelvin. The noise power is given by

$$N = kT_A B \tag{5.1}$$

where B is the bandwidth and k is Boltzmann's constant,

$$k = 1.38 \times 10^{-23} \text{ J/K}$$

Static or atmospheric noise is due to a flash of lightning somewhere in the world. The lightning generates an impulse noise that has the greatest magnitude at 10 kHz and is negligible at frequencies greater than 20 MHz.

Galactic noise is produced by radiation from distant stars. It has a maximum value at about 20 MHz and is negligible above 500 MHz.

Man-made noise includes many different sources. For example, when electric current is switched on or off, voltage spikes will be generated. These transient spikes occur in electronic or mechanical switches, vehicle ignition systems, light switches, motors, and so on. Electromagnetic radiation from communication systems, broadcast systems, radar, and power lines is everywhere, and the undesired signals can be picked up by a receiver. The interference is always present and could be severe in urban areas.

In addition to the noise picked up by the antenna, the receiver itself adds further noise to the signal from its amplifier, filter, mixer, and detector stages. The quality of the output signal from the receiver for its intended purpose is expressed in terms of its signal-to-noise ratio (SNR):

$$ \text{SNR} = \frac{\text{wanted signal power}}{\text{unwanted noise power}} \tag{5.2} $$

A tangential detectable signal is defined as SNR = 3 dB (or a factor of 2). For a mobile radio-telephone system, SNR > 15 dB is required from the receiver output. In a radar system, the higher SNR corresponds to a higher probability of detection and a lower false-alarm rate. An SNR of 16 dB gives a probability detection of 99.99% and a probability of false-alarm rate of 10^{-6} [2].

The noise that occurs in a receiver acts to mask weak signals and to limit the ultimate sensitivity of the receiver. In order for a signal to be detected, it should have a strength much greater than the noise floor of the system. Noise sources in thermionic and solid-state devices may be divided into three major types.

1. *Thermal, Johnson, or Nyquist Noise.* This noise is caused by the random fluctuations produced by the thermal agitation of the bound charges. The *rms* value of the thermal resistance noise voltage of V_n over a frequency range B is given by

$$ V_n^2 = 4kTBR \tag{5.3} $$

where k = Boltzman constant = 1.38×10^{-23} J/K
T = resistor absolute temperature, K
B = bandwidth, Hz
R = resistance, Ω

From Eq. (5.3), the noise power can be found to exist in a given bandwidth regardless of the center frequency. The distribution of the same noise-per-unit bandwidth everywhere is called white noise.

2. *Shot Noise.* The fluctuations in the number of electrons emitted from the source constitute the shot noise. Shot noise occurs in tubes or solid-state devices.

3. *Flicker, or* $1/f$, *Noise.* A large number of physical phenomena, such as mobility fluctuations, electromagnetic radiation, and quantum noise [3], exhibit a noise power that varies inversely with frequency. The $1/f$ noise is important from 1 Hz to 1 MHz. Beyond 1 MHz, the thermal noise is more noticeable.

5.4 RECEIVER NOISE FIGURE AND EQUIVALENT NOISE TEMPERATURE

Noise figure is a figure of merit quantitatively specifying how noisy a component or system is. The noise figure of a system depends on a number of factors such as losses in the circuit, the solid-state devices, bias applied, and amplification. The noise factor of a two-port network is defined as

$$F = \frac{\text{SNR at input}}{\text{SNR at output}} = \frac{S_i/N_i}{S_o/N_o} \tag{5.4}$$

The noise figure is simply the noise factor converted in decibel notation.

Figure 5.4 shows the two-port network with a gain (or loss) G. We have

$$S_o = GS_i \tag{5.5}$$

Note that $N_o \neq GN_i$; instead, the output noise $N_o = GN_i +$ noise generated by the network. The noise added by the network is

$$N_n = N_o - GN_i \quad \text{(W)} \tag{5.6}$$

Substituting (5.5) into (5.4), we have

$$F = \frac{S_i/N_i}{GS_i/N_o} = \frac{N_o}{GN_i} \tag{5.7}$$

Therefore,

$$N_o = FGN_i \quad \text{(W)} \tag{5.8}$$

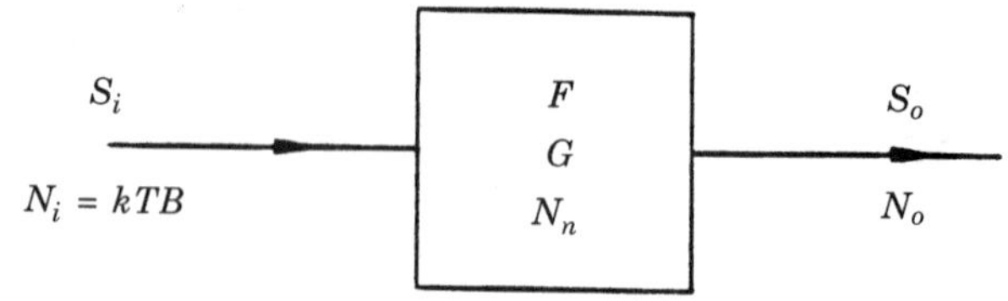

FIGURE 5.4 Two-port network with gain G and added noise power N_n.

Equation (5.8) implies that the input noise N_i (in decibels) is raised by the noise figure F (in decibels) and the gain (in decibels).

Since the noise figure of a component should be independent of the input noise, F is based on a standard input noise source N_i at room temperature in a bandwidth B, where

$$N_i = kT_0B \quad \text{(W)} \tag{5.9}$$

where k is the Boltzmann constant, $T_0 = 290$ K (room temperature), and B is the bandwidth. Then, Eq. (5.7) becomes

$$F = \frac{N_o}{GkT_0B} \tag{5.10}$$

For a cascaded circuit with n elements as shown in Fig. 5.5, the overall noise factor can be found from the noise factors and gains of the individual elements [4]:

$$F = F_1 + \frac{F_2 - 1}{G_1} + \frac{F_3 - 1}{G_1G_2} + \cdots + \frac{F_n - 1}{G_1G_2 \cdots G_{n-1}} \tag{5.11}$$

Equation (5.11) allows for the calculation of the noise figure of a general cascaded system. From Eq. (5.11), it is clear that the gain and noise figure in the first stage are critical in achieving a low overall noise figure. It is very desirable to have a low noise figure and high gain in the first stage. To use Eq. (5.11), all F's and G's are in ratio. For a passive component with loss L in ratio, we will have $G = 1/L$ and $F = L$ [4].

Example 5.1   For the two-element cascaded circuit shown in Fig. 5.6, prove that the overall noise factor

$$F = F_1 + \frac{F_2 - 1}{G_1}$$

Solution   From Eq. (5.10)

$$N_o = F_{12}G_{12}kT_0B \qquad N_{o1} = F_1G_1kT_0B$$

From Eqs. (5.6) and (5.8)

$$N_{n2} = (F_2 - 1)G_2kT_0B$$

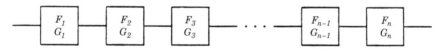

FIGURE 5.5   Cascaded circuit with n networks.

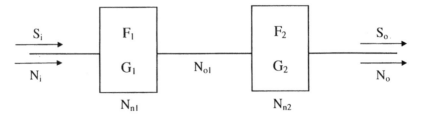

FIGURE 5.6   Two-element cascaded circuit.

From Eq. (5.6)

$$N_o = N_{o1}G_2 + N_{n2}$$

Substituting the first three equations into the last equation leads to

$$N_o = F_1 G_1 G_2 kT_0 B + (F_2 - 1)G_2 kT_0 B$$
$$= F_{12} G_{12} kT_0 B$$

Overall,

$$F = F_{12} = \frac{F_1 G_1 G_2 kT_0 B}{G_1 G_2 kT_0 B} + \frac{(F_2 - 1)G_2 kT_0 B}{G_1 G_2 kT_0 B}$$
$$= F_1 + \frac{F_2 - 1}{G_1}$$

The proof can be generalized to n elements.   ∎

Example 5.2   Calculate the overall gain and noise figure for the system shown in Fig. 5.7.

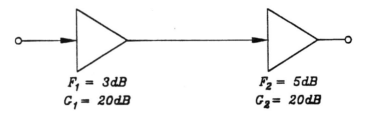

FIGURE 5.7   Cascaded amplifiers.

Solution

$$F_1 = 3 \text{ dB} = 2 \qquad F_2 = 5 \text{ dB} = 3.162$$
$$G_1 = 20 \text{ dB} = 100 \qquad G_2 = 20 \text{ dB} = 100$$
$$G = G_1 G_2 = 10{,}000 = 40 \text{ dB}$$
$$F = F_1 + \frac{F_2 - 1}{G_1} = 2 + \frac{3.162 - 1}{100}$$
$$= 2 + 0.0216 = 2.0216 = 3.06 \text{ dB}. \qquad \blacksquare$$

Note that $F \approx F_1$ due to the high gain in the first stage. The first-stage amplifier noise figure dominates the overall noise figure. One would like to select the first-stage RF amplifier with a low noise figure and a high gain to ensure the low noise figure for the overall system.

The equivalent noise temperature is defined as

$$T_e = (F - 1)T_0 \qquad (5.12)$$

where $T_0 = 290$ K (room temperature) and F in ratio. Therefore,

$$F = 1 + \frac{T_e}{T_0} \qquad (5.13)$$

Note that T_e is not the physical temperature. From Eq. (5.12), the corresponding T_e for each F is given as follows:

F (dB)	3	2.28	1.29	0.82	0.29
T_e (K)	290	200	100	60	20

For a cascaded circuit shown as Fig. 5.8, Eq. (5.11) can be rewritten as

$$T_e = T_{e1} + \frac{T_{e2}}{G_1} + \frac{T_{e3}}{G_1 G_2} + \cdots + \frac{T_{en}}{G_1 G_2 \cdots G_{n-1}} \qquad (5.14)$$

where T_e is the overall equivalent noise temperature in kelvin.

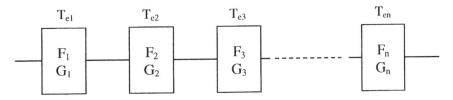

FIGURE 5.8   Noise temperature for a cascaded circuit.

The noise temperature is useful for noise factor calculations involving an antenna. For example, if an antenna noise temperature is T_A, the overall system noise temperature including the antenna is

$$T_S = T_A + T_e \tag{5.15}$$

where T_e is the overall cascaded circuit noise temperature.

As pointed out earlier in Section 5.3, the antenna noise temperature is approximately equal to 290 K for an antenna pointing to earth. The antenna noise temperature could be very low (a few kelvin) for an antenna pointing to the sky.

5.5 COMPRESSION POINTS, MINIMUM DETECTABLE SIGNAL, AND DYNAMIC RANGE

In a mixer, an amplifier, or a receiver, operation is normally in a region where the output power is linearly proportional to the input power. The proportionality constant is the conversion loss or gain. This region is called the dynamic range, as shown in Fig. 5.9. For an amplifier, the curve shown in Fig. 5.9 is for the fundamental signals. For a mixer or receiver, the curve is for the IF signals. If the input power is above this range, the output starts to saturate. If the input power is below this range, the noise dominates. The dynamic range is defined as the range between the 1-dB compression point and the minimum detectable signal (MDS). The range could be specified in terms of input power (as shown in Fig. 5.9) or output power. For a mixer, amplifier, or receiver system, we would like to have a high dynamic range so the system can operate over a wide range of input power levels.

The noise floor due to a matched resistor load is

$$N_i = kTB \tag{5.16}$$

where k is the Boltzmann constant. If we assume room temperature (290 K) and 1 MHz bandwidth, we have

$$
\begin{aligned}
N_i &= 10 \log kTB = 10 \log(4 \times 10^{-12} \text{ mW}) \\
&= -114 \text{ dBm}
\end{aligned}
\tag{5.17}
$$

The MDS is defined as 3 dB above the noise floor and is given by

$$
\begin{aligned}
\text{MDS} &= -114 \text{ dBm} + 3 \text{ dB} \\
&= -111 \text{ dBm}
\end{aligned}
\tag{5.18}
$$

Therefore, MDS is -111 dBm (or 7.94×10^{-12} mW) in a megahertz bandwidth at room temperature.

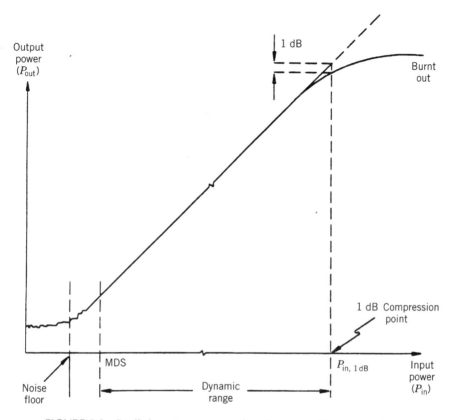

FIGURE 5.9   Realistic system response for mixers, amplifiers, or receivers.

The 1-dB compression point is shown in Fig. 5.9. Consider an example for a mixer. Beginning at the low end of the dynamic range, just enough RF power is fed into the mixer to cause the IF signal to be barely discernible above the noise. Increasing the RF input power causes the IF output power to increase decibel for decibel of input power; this continues until the RF input power reaches a level at which the IF output power begins to roll off, causing an increase in conversion loss. The input power level at which the conversion loss increases by 1 dB, called the 1-dB compression point, is generally taken to be the top limit of the dynamic range. Beyond this range, the conversion loss is higher, and the input RF power not converted into the desired IF output power is converted into heat and higher order intermodulation products.

In the linear region for an amplifier, a mixer, or a receiver,

$$P_{in} = P_{out} - G \tag{5.19}$$

where G is the gain of the receiver or amplifier, $G = -L_c$ for a lossy mixer with a conversion loss L_c (in decibels).

The input signal power in dBm that produces a 1-dB gain in compression is shown in Fig. 5.9 and given by

$$P_{\text{in, 1dB}} = P_{\text{out, 1dB}} - G + 1 \text{ dB} \tag{5.20}$$

for an amplifier or a receiver with gain.
   For a mixer with conversion loss,

$$P_{\text{in, 1dB}} = P_{\text{out, 1dB}} + L_c + 1 \text{ dB} \tag{5.21}$$

or one can use Eq. (5.20) with a negative gain. Note that $P_{\text{in, 1dB}}$ and $P_{\text{out, 1dB}}$ are in dBm, and gain and L_c are in decibels. Here $P_{\text{out, 1dB}}$ is the output power at the 1-dB compression point, and $P_{\text{in, 1dB}}$ is the input power at the 1-dB compression point. Although the 1-dB compression points are most commonly used, 3-dB compression points and 10-dB compression points are also used in some system specifications.
   From the 1-dB compression point, gain, bandwidth, and noise figure, the dynamic range (DR) of a mixer, an amplifier, or a receiver can be calculated. The DR can be defined as the difference between the input signal level that causes a 1-dB compression gain and the minimum input signal level that can be detected above the noise level:

$$DR = P_{\text{in, 1dB}} - MDS \tag{5.22}$$

Note that $P_{\text{in, 1dB}}$ and MDS are in dBm and DR in decibels.

Example 5.3   A receiver operating at room temperature has a noise figure of 5.5 dB and a bandwidth of 2 GHz. The input 1-dB compression point is +10 dBm. Calculate the minimum detectable signal and dynamic range.

Solution

$$F = 5.5 \text{ dB} = 3.6 \qquad B = 2 \times 10^9 \text{ Hz}$$

$$MDS = 10 \log kTBF + 3 \text{ dB}$$

$$= 10 \log(1.38 \times 10^{-23} \times 290 \times 2 \times 10^9 \times 3.6) + 3$$

$$= -102.5 \text{ dBW} = -72.5 \text{ dBm}$$

$$DR = P_{\text{in, 1dB}} - MDS = 10 \text{ dBm} - (-72.5 \text{ dBm}) = 82.5 \text{ dB} \qquad \blacksquare$$

5.6   THIRD-ORDER INTERCEPT POINT AND INTERMODULATION

When two or more signals at frequencies f_1 and f_2 are applied to a nonlinear device, they generate IM products according to $mf_1 \pm nf_2$ (where $m, n = 0, 1, 2, \ldots$). These may be the second-order $f_1 \pm f_2$ products, third-order $2f_1 \pm f_2$, $2f_2 \pm f_1$ products, and so on. The two-tone third-order IM products are of primary interest since they tend to have frequencies that are within the passband of the first IF stage.

Consider a mixer or receiver as shown in Fig. 5.10, where f_{IF1} and f_{IF2} are the desired IF outputs. In addition, the third-order IM (IM3) products f_{IM1} and f_{IM2} also appear at the output port. The third-order intermodulation (IM3) products are generated from f_1 and f_2 mixing with one another and then beating with the mixer's LO according to the expressions

$$(2f_1 - f_2) - f_{LO} = f_{IM1} \tag{5.23a}$$

$$(2f_2 - f_1) - f_{LO} = f_{IM2} \tag{5.23b}$$

where f_{IM1} and f_{IM2} are shown in Fig. 5.11 with IF products for f_{IF1} and f_{IF2} generated by the mixer or receiver:

$$f_1 - f_{LO} = f_{IF1} \tag{5.24}$$

$$f_2 - f_{LO} = f_{IF2} \tag{5.25}$$

Note that the frequency separation is

$$\Delta = f_1 - f_2 = f_{IM1} - f_{IF1} = f_{IF1} - f_{IF2} = f_{IF2} = f_{IM2} \tag{5.26}$$

These intermodulation products are usually of primary interest because of their relatively large magnitude and because they are difficult to filter from the desired mixer outputs (f_{IF1} and f_{IF2}) if Δ is small.

The intercept point, measured in dBm, is a figure of merit for intermodulation product suppression. A high intercept point indicates a high suppression of undesired intermodulation products. The third-order intercept point (IP3 or TOI) is the theoretical point where the desired signal and the third-order distortion have equal magnitudes. The TOI is an important measure of the system's linearity. A

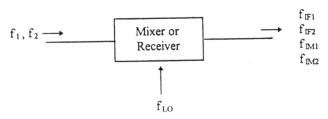

FIGURE 5.10   Signals generated from two RF signals.

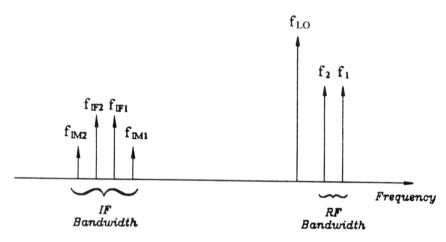

FIGURE 5.11   Intermodulation products.

convenient method for determining the two-tone third-order performance of a mixer is the TOI measurement. Typical curves for a mixer are shown in Fig. 5.12. It can be seen that the 1-dB compression point occurs at the input power of +8 dBm. The TOI point occurs at the input power of +16 dBm, and the mixer will suppress third-order products over 55 dB with both signals at −10 dBm. With both input signals at 0 dBm, the third-order products are suppressed over 35 dB, or one can say that IM3 products are 35 dB below the IF signals. The mixer operates with the LO at 57 GHz and the RF swept from 60 to 63 GHz. The conversion loss is less than 6.5 dB.

In the linear region, for the IF signals, the output power is increased by 1 dB if the input power is increased by 1 dB. The IM3 products are increased by 3 dB for a 1-dB increase in P_{in}. The slope of the curve for the IM3 products is 3 : 1.

For a cascaded circuit, the following procedure can be used to calculate the overall system intercept point [6] (see Example 5.5):

1. Transfer all input intercept points to system input, subtracting gains and adding losses decibel for decibel.
2. Convert intercept points to powers (dBm to milliwatts). We have IP_1, IP_2, ..., IP_N for N elements.
3. Assuming all input intercept points are independent and uncorrelated, add powers in "parallel":

$$IP3_{input} = \left(\frac{1}{IP_1} + \frac{1}{IP_2} + \cdots + \frac{1}{IP_N} \right)^{-1} \quad (mW) \qquad (5.27)$$

4. Convert $IP3_{input}$ from power (milliwatts) to dBm.

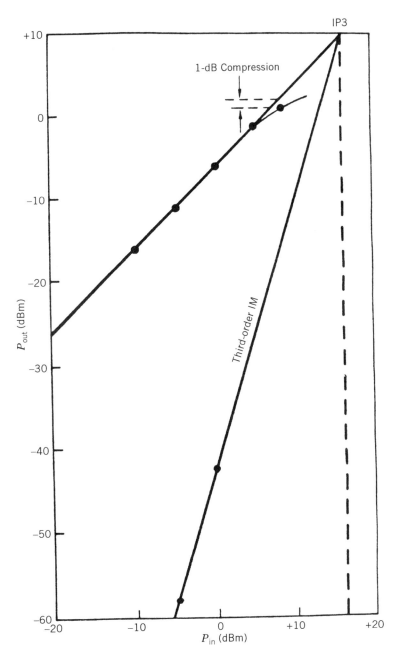

FIGURE 5.12   Intercept point and 1-dB compression point measurement of a V-band crossbar stripline mixer. (From reference [5], with permission from IEEE.)

Example 5.4   When two tones of -10 dBm power level are applied to an amplifier, the level of the IM3 is -50 dBm. The amplifier has a gain of 10 dB. Calculate the IM3 output power when the power level of the two-tone is -20 dBm. Also, indicate the IM3 power as decibels down from the wanted signal.

Solution $P_{in} = -20\text{dBm}$

As shown in Fig. 5.13,

$$\text{IM3 power} = (-50 \text{ dBm}) + 3 \times [-20 \text{ dBm} - (-10 \text{ dBm})]$$
$$= -50 \text{ dBm} - 30 \text{ dBm} = -80 \text{ dBm}$$

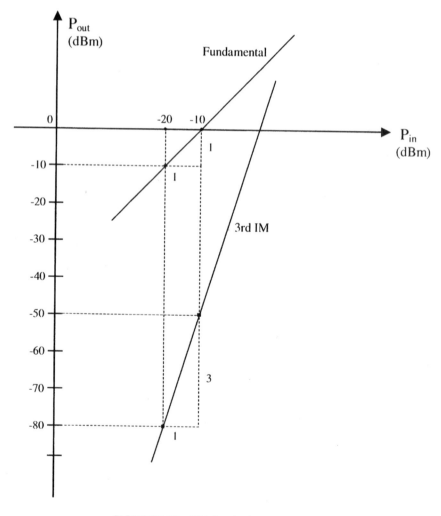

FIGURE 5.13   Third-order intermodulation.

Then

$$\text{Wanted signal at } P_{in} = -20 \text{ dBm has a power level}$$
$$= -20 \text{ dBm} + \text{gain} = -10 \text{ dBm}$$
$$\text{Difference between wanted signal and IM3}$$
$$= -10 \text{ dBm} - (-80 \text{ dBm}) = 70 \text{ dB down} \qquad ■$$

Example 5.5 A receiver is shown in Fig. 5.14. Calculate the overall input IP3 in dBm.

Solution   Transfer all intercept points to system input; the results are shown in Fig. 5.14. The overall input IP3 is given by

$$IP3 = 10 \log \left(\frac{1}{IP_1} + \frac{1}{IP_2} + \frac{1}{IP_3} + \frac{1}{IP_4} + \frac{1}{IP_5} \right)^{-1}$$

$$= 10 \log \left(\frac{1}{\infty} + \frac{1}{15.85} + \frac{1}{\infty} + \frac{1}{19.95} + \frac{1}{100} \right)^{-1}$$

$$= 10 \log 8.12 \text{ mW} = 9.10 \text{ dBm} \qquad ■$$

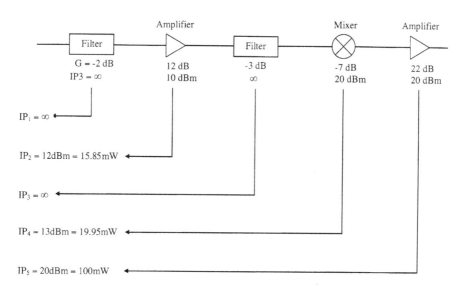

FIGURE 5.14   Receiver and its input intercept point.

5.7 SPURIOUS RESPONSES

Any undesirable signals are spurious signals. The spurious signals could produce demodulated output in the receiver if they are at a sufficiently high level. This is especially troublesome in a wide-band receiver. The spurious signals include the harmonics, intermodulation products, and interferences.

The mixer is a nonlinear device. It generates many signals according to $\pm m f_{RF} \pm n f_{LO}$, where $m = 0, 1, 2, \ldots$ and $n = 0, 1, 2, \ldots$, although a filter is used at the mixer output to allow only f_{IF} to pass. Other low-level signals will also appear at the output. If $m = 0$, a whole family of spurious responses of LO harmonics or $n f_{LO}$ spurs are generated.

Any RF frequency that satisfies the following equation can generate spurious responses in a mixer:

$$m f_{RF} - n f_{LO} = \pm f_{IF} \tag{5.28}$$

where f_{IF} is the desired IF frequency.

Solving (5.28) for f_{RF}, each (m, n) pair will give two possible spurious frequencies due to the two RF frequencies:

$$f_{RF1} = \frac{n f_{LO} - f_{IF}}{m} \tag{5.29}$$

$$f_{RF2} = \frac{n f_{LO} + f_{IF}}{m} \tag{5.30}$$

The RF frequencies of f_{RF1} and f_{RF2} will generate spurious responses.

5.8 SPURIOUS-FREE DYNAMIC RANGE

Another definition of dynamic range is the "spurious-free" region that characterizes the receiver with more than one signal applied to the input. For the case of input signals at equal levels, the spurious-free dynamic range SFDR or DR_{sf} is given by

$$DR_{sf} = \tfrac{2}{3}(IP3 - MDS) \tag{5.31}$$

where IP3 is the input power at the third-order, two-tone intercept point in dBm and MDS is the input minimum detectable signal.

Equation (5.31) can be proved in the following: From Fig. 5.15, one has

$$BD = \tfrac{1}{3}CD \qquad EB = AB$$

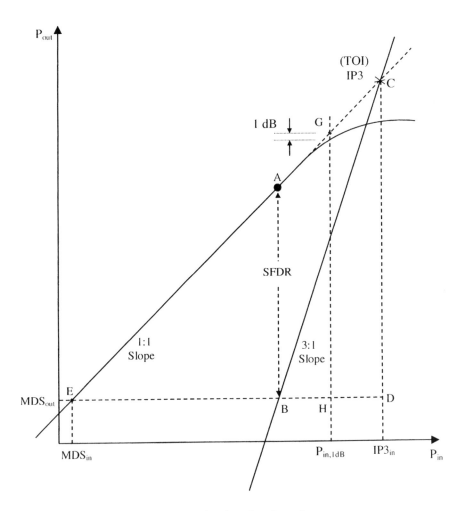

FIGURE 5.15   Spurious-free dynamic range.

From the triangle CED, we have

$$CD = ED = EB + BD = AB + \tfrac{1}{3}CD$$

Therefore,

$$AB = \tfrac{2}{3}CD = \tfrac{2}{3}(\mathrm{IP3}_{\mathrm{out}} - \mathrm{MDS}_{\mathrm{out}})$$

or since $CD = ED$,

$$\mathrm{DR}_{\mathrm{sf}} = AB = \tfrac{2}{3}ED = \tfrac{2}{3}(\mathrm{IP3}_{\mathrm{in}} - \mathrm{MDS}_{\mathrm{in}})$$

and *AB* is the spurious-free dynamic range. Note that *GH* is the dynamic range, which is defined by

$$DR = GH = EH = P_{\text{in, 1dB}} - \text{MDS}_{\text{in}}$$

The IP3_{in} and IP3_{out} differ by the gain (or loss) of the system. Similarly, MDS_{in} differs from MDS_{out} by the gain (or loss) of the system.

PROBLEMS

5.1   Calculate the overall noise figure and gain in decibels for the system (at room temperature, 290 K) shown in Fig. P5.1.

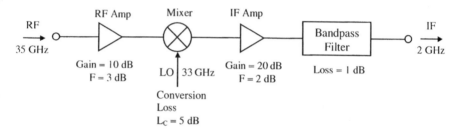

FIGURE P5.1

5.2   The receiver system shown in Fig. P5.2 is used for communication systems. The 1-dB compression point occurs at the output IF power of +20 dBm. At room temperature, calculate (a) the overall system gain or loss in decibels, (b) the overall noise figure in decibels, (c) the minimum detectable signal in milliwatts at the input RF port, and (d) the dynamic range in decibels.

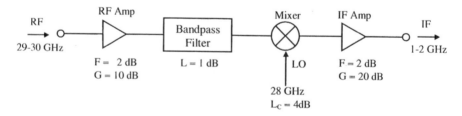

FIGURE P5.2

5.3   A receiver operating at room temperature is shown in Fig. P5.3. The receiver input 1-dB compression point is +10 dBm. Determine (a) the overall gain in decibels, (b) the overall noise figure in decibels, and (c) the dynamic range in decibels.

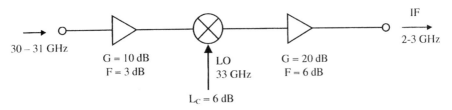

FIGURE P5.3

5.4   The receiver system shown in Fig. P5.4 has the following parameters: $P_{in, 1dB} = +10$ dBm, $IP3_{in} = 20$ dBm. The receiver is operating at room temperature. Determine (a) the noise figure in decibels, (b) the dynamic range in decibels, (c) the output SNR ratio for an input SNR ratio of 10 dB, and (d) the output power level in dBm at the 1-dB compression point.

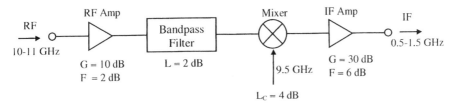

FIGURE P5.4

5.5   Calculate the overall system noise temperature and its equivalent noise figure in decibels for the system shown in Fig. P5.5.

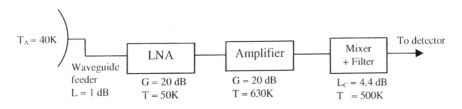

FIGURE P5.5

5.6   When two 0-dBm tones are applied to a mixer, the level of the IM3 is -60 dBm. The mixer has a conversion loss of 6 dB. Assume that the 1-dB compression point has input power generated greater than $+13$ dBm. (a) Indicate the IM3 power as how many decibels down from the wanted signal. (b) Calculate the IM3 output power when the level of the two tones is -10 dBm, and indicate the IM3 power as decibels down from the wanted signal. (c) Repeat part (b) for the two-tone level of $+10$ dBm.

5.7   At an input signal power level of -10 dBm, the output wanted signal from a receiver is 50 dB above the IM3 products (i.e., 50 dB suppression of the IM3 products). If the input signal level is increased to 0 dBm, what is the suppression level for the IM3 products?

5.8   When two tones of −20 dBm power level are incident to an amplifier, the level of the IM3 is −80 dBm. The amplifier has a gain of 10 dB. Calculate the IM3 output power when the power level of the two tones is −10 dBm. Also, indicate the IM3 power as decibels down from the wanted signal.

5.9   Calculate the overall system IP3 power level for the system shown in Fig. P5.9.

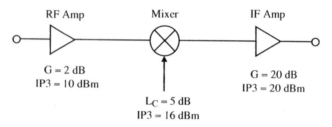

FIGURE P5.9

5.10   For the system shown in Fig. P5.10, calculate (a) the overall system gain in decibels, (b) the overall noise figure in decibels, (c) the equivalent noise temperature in kelvin, (d) the minimum detectable signal (MDS) in dBm at input port, and (e) the input IP3 power level in dBm. The individual component system parameters are given in the figure, and the system is operating at room temperature (290 K).

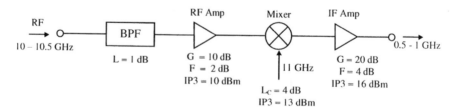

FIGURE P5.10

5.11   A radio receiver operating at room temperature has the block diagram shown in Fig. P5.11. Calculate (a) the overall gain/loss in decibels, (b) the overall noise figure in decibels, and (c) the input IP3 power level in dBm. (d) If the input signal power is 0.1 mW and the SNR is 20 dB, what are the output power level and the SNR?

5.12   In the system shown in Fig. P5.12, determine (a) the overall gain in decibels, (b) the overall noise figure in decibels, and (c) the overall intercept point power level in dBm at the input.

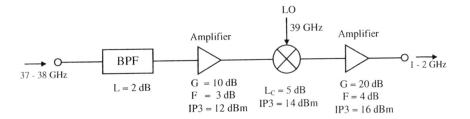

FIGURE P5.11

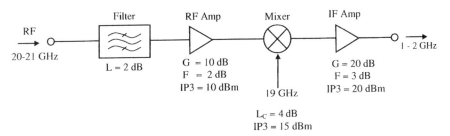

FIGURE P5.12

REFERENCES

1. T. Stetzler et al., "A 2.7 V to 4.5 V Single Chip GSM Transceiver RF Integrated Circuit," *1995 IEEE International Solid-State Circuits Conference*, pp. 150–151, 1995.
2. M. L. Skolnik, *Introduction to Radar Systems*, 2nd ed., McGraw-Hill, New York, 1980.
3. S. Yugvesson, *Microwave Semiconductor Devices*, Kluwer Academic, The Netherlands, 1991, Ch. 8.
4. K. Chang, *Microwave Solid-State Circuits and Applications*, John Wiley & Sons, New York, 1994.
5. K. Chang, K. Louie, A. J. Grote, R. S. Tahim, M. J. Mlinar, G. M. Hayashibara, and C. Sun, "V-Band Low-Noise Integrated Circuit Receiver," *IEEE Trans. Microwave Theory Tech.*, Vol. MTT-31, pp. 146–154, 1983.
6. P. Vizmuller, *RF Design Guide*, Artech House, Boston, 1995.

CHAPTER SIX

Transmitter and Oscillator Systems

6.1 TRANSMITTER PARAMETERS

A transmitter is an important subsystem in a wireless system. In any active wireless system, a signal will be generated and transmitted through an antenna. The signal's generating system is called a transmitter. The specifications for a transmitter depend on the applications. For long-distance transmission, high power and low noise are important. For space or battery operating systems, high efficiency is essential. For communication systems, low noise and good stability are required. A transmitter can be combined with a receiver to form a transceiver. In this case, a duplexer is used to separate the transmitting and receiving signals. The duplexer could be a switch, a circulator, or a diplexer, as described in Chapter 4.

A transmitter generally consists of an oscillator, a modulator, an upconverter, filters, and power amplifiers. A simple transmitter could have only an oscillator, and a complicated one would include a phase-locked oscillator or synthesizer and the above components. Figure 6.1 shows a typical transmitter block diagram. The information will modulate the oscillator through AM, FM, phase modulation (PM), or digital modulation. The output signal could be upconverted to a higher frequency. The power amplifiers are used to increase the output power before it is transmitted by an antenna. To have a low phase noise, the oscillator or local oscillator can be phase locked to a low-frequency crystal oscillator. The oscillator could also be replaced by a frequency synthesizer that derives its frequencies from an accurate high-stability crystal oscillator source. The following transmitter characteristics are of interest:

1. *Power output and operating frequency*: the output RF power level generated by a transmitter at a certain frequency or frequency range.

Oscillator

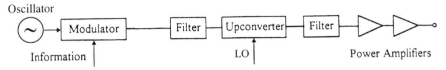

FIGURE 6.1   Transmitter system.

2. *Efficiency*: the DC-to-RF conversion efficiency of the transmitter.
3. *Power output variation*: the output power level variation over the frequency range of operation.
4. *Frequency tuning range*: the frequency tuning range due to mechanical or electronic tuning.
5. *Stability*: the ability of an oscillator/transmitter to return to the original operating point after experiencing a slight thermal, electrical, or mechanical disturbance.
6. *Circuit quality (Q) factor*: the loaded and unloaded Q-factor of the oscillator's resonant circuit.
7. *Noise*: the AM, FM, and phase noise. Amplitude-modulated noise is the unwanted amplitude variation of the output signal, frequency-modulated noise is the unwanted frequency variations, and phase noise is the unwanted phase variations.
8. *Spurious signals*: output signals at frequencies other than the desired carrier.
9. *Frequency variations*: frequency jumping, pulling, and pushing. Frequency jumping is a discontinuous change in oscillator frequency due to nonlinearities in the device impedance. Frequency pulling is the change in oscillator frequency versus a specified load mismatch over 360° of phase variation. Frequency pushing is the change in oscillator frequency versus DC bias point variation.
10. *Post-tuning drift*: frequency and power drift of a steady-state oscillator due to heating of a solid-state device.

Some of these characteristics can be found in an example given in Table 6.1.

6.2   TRANSMITTER NOISE

Since the oscillator is a nonlinear device, the noise voltages and currents generated in an oscillator are modulating the signal produced by the oscillator. Figure 6.2 shows the ideal signal and the signal modulated by the noise. The noise can be classified as an AM noise, FM noise, and phase noise.

Amplitude-modulated noise causes the amplitude variations of the output signal. Frequency-modulated or phase noise is indicated in Fig. 6.2*b* by the spreading of the

TABLE 6.1 Typical Commercial Voltage-Controlled Oscillator (VCO) Specifications

Frequency (f_0)	35 GHz
Power (P_0)	250 mW
Bias pushing range (typical)	50 MHz/V
Varactor tuning range	±250 MHz
Frequency drift over temperature	−2 MHz/°C
Power drop over temperature	−0.03 dB/°C
Q_{ext}	800–1000
Harmonics level	−200 dBc minimum
Modulation bandwidth	DC − 50 MHz
Modulation sensitivity (MHz/V)	25–50
FM noise at 100-kHz offset	−90 dBc/kHz or −120 dBc/Hz
AM noise at 100-kHz offset	−155 dBc/kHz or −185 dBc/Hz

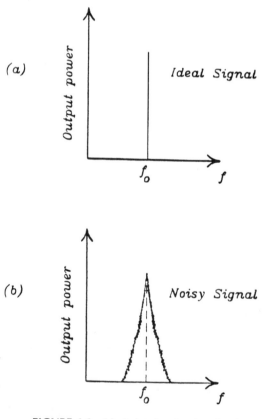

FIGURE 6.2 Ideal signal and noisy signal.

frequency spectrum. A ratio of single-sideband noise power normalized in 1-Hz bandwidth to the carrier power is defined as

$$\mathscr{L}(f_m) = \frac{\text{noise power in 1-Hz bandwidth at } f_m \text{ offset from carrier}}{\text{carrier signal power}}$$

$$= \frac{N}{C} \tag{6.1}$$

As shown in Fig. 6.3, $\mathscr{L}(f_m)$ is the difference of power between the carrier at f_0 and the noise at $f_0 + f_m$. The power is plotted in the decibel scale, and the unit of $\mathscr{L}(f_m)$ is in decibels below the carrier power (dBc) per hertz. The FM noise is normally given as the number of decibels below carrier amplitude at a frequency f_m that is offset from the carrier. Figure 6.4 shows a typical phase noise measurement from a Watkins–Johnson dielectric resonator oscillator (DRO) [1]. The phase noise is 70 dBc/Hz at 1 kHz offset from the carrier and 120 dBc/Hz at 100 KHz offset from the carrier. Here dBc/Hz means decibels below carrier over a bandwidth of 1 Hz.

It should be mentioned that the bulk of oscillator noise close to the carrier is the phase or FM noise. The noise represents the phase jitter or the short-term stability of the oscillator. The oscillator power is not concentrated at a single frequency but is rather distributed around it. The spectral distributions on the opposite sides of the carrier are known as noise sidebands. To minimize the FM noise, one can use a high-

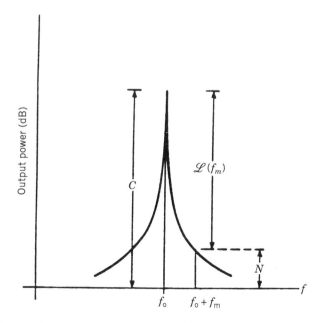

FIGURE 6.3 Oscillator output power spectrum. This spectrum can be seen from the screen of a spectrum analyzer.

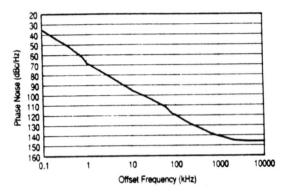

FIGURE 6.4 Phase noise measurement for a WJ VC1001 DRO [1]. (Courtesy of Watkins-Johnson.)

Q resonant circuit, a low-noise active device, a phase-locked loop, or avoid the operation in a region of saturation.

Many methods can be used to measure the FM or phase noise [2–4]. These methods include the spectrum analyzer method, the two-oscillator method, the single-oscillator method, the delay-line discriminator method, and the cavity discriminator method.

6.3 FREQUENCY STABILITY AND SPURIOUS SIGNALS

Slight electrical, thermal, or mechanical disturbances can cause an oscillator to change operating frequency. The disturbance may cause the oscillation to cease since it could change the device impedance such that the oscillating conditions described in Chapter 4 are no longer satisfied.

Stability is a measure that describes an oscillator's ability to return to its steady-state operating point. The temperature stability can be specified in three different ways. For example, at 10 GHz, an oscillator or a transmitter has the following temperature stability specifications: ± 10 KHz/$^\circ$C, or ± 800 KHz over -30°C to $+50^\circ$C, or ± 1 ppm/$^\circ$C, where ppm stands for parts per million. At 10 GHz, ± 1 ppm/$^\circ$C is equivalent to ± 10 KHz/$^\circ$C. This can be seen from the following:

$$\pm 1 \text{ ppm/}^\circ\text{C} \times 10 \text{ GHz} = \pm 1 \times 10^{-6} \times 10 \times 10^9 \text{ Hz/}^\circ\text{C}$$
$$= \pm 10 \text{ KHz/}^\circ\text{C}$$

A typical wireless communication system requires a stability range from 0.5 to 5 ppm/$^\circ$C and a phase noise range from -80 to -120 dBc/Hz.

Frequency variations could be due to other problems such as frequency jumping, pulling, and pushing, as described in Section 6.1. Post-tuning drift can also change the desired operating frequency.

The transmitter with good stability and low noise is important for wireless communication applications. To improve the stability, one can use (1) high-Q circuits to build the oscillators (examples are waveguide cavities, dielectric resonators, or superconducting resonators/cavities); (2) temperature compensation circuits; or (3) phase-locked oscillators or frequency synthesizers, which will be discussed later in this chapter.

For an oscillator, spurious signals are the undesired signals at frequencies other than the desired oscillation signal. These include the harmonics and bias oscillations. The harmonic signals have frequencies that are integer multiples of the oscillating frequency. If the oscillating frequency is f_0, the second harmonic is $2f_0$, and the third harmonic is $3f_0$, and so on. As shown in Fig. 6.5, the power levels of harmonics are generally well below the fundamental frequency power. A specification for harmonic power is given by the number of decibels below carrier. For example, second-harmonic output is -30 dBc and third-harmonic output is -60 dBc. For a complicated transmitter with upconverters and power amplifiers, many other spurious signals could exist at the output due to the nonlinearity of these components. The nonlinearity will cause two signals to generate many mixing and intermodulation products.

6.4 FREQUENCY TUNING, OUTPUT POWER, AND EFFICIENCY

The oscillating frequency is determined by the resonant frequency of the overall oscillator circuit. At resonance, the total reactance (or susceptance) equals zero. Consider a simplified circuit shown in Fig. 6.6, where Z_D is the active device impedance and Z_C is the external circuit impedance. The oscillating (or resonant) frequency is the frequency such that

$$\mathrm{Im}(Z_D) + \mathrm{Im}(Z_C) = 0 \tag{6.2}$$

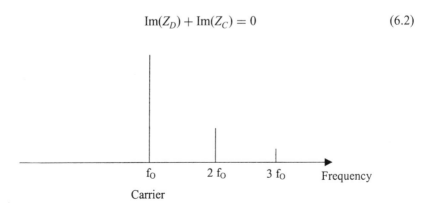

FIGURE 6.5 Oscillating frequency and its harmonics.

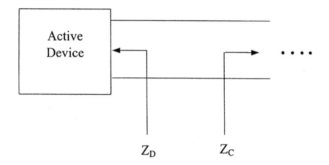

FIGURE 6.6 Simplified oscillator circuit.

where Im stands for the imaginary part. The circuit impedance is a function of frequency only, and the device impedance is a function of frequency (f), bias current (I_0), generated RF current (I_{RF}), and temperature (T). Therefore, at the resonant frequency, we have

$$\text{Im}[Z_D(f, I_0, I_{RF}, T)] + \text{Im}[Z_C(f)] = 0 \tag{6.3}$$

Electronic frequency tuning can be accomplished by bias tuning or varactor tuning. The bias tuning will change I_0 and thus change Z_D, resulting in a new oscillating frequency. The varactor tuning (as shown in Fig. 6.7 as an example) will change $C(V)$ and thus change Z_C, resulting in a new oscillating frequency. The frequency tuning is useful for frequency modulation in radar or communication systems. For example, a 10-GHz voltage-controlled oscillator (VCO) could have a modulation sensitivity of 25 MHz/V and a tuning range of ±100 MHz by varying the bias voltage to a varactor.

For most systems, a constant output power is desirable. Power output could vary due to temperature, bias, frequency tuning, and environment. A specification for power variation can be written as 30 dBm $\pm$ 0.5 dB, as an example.

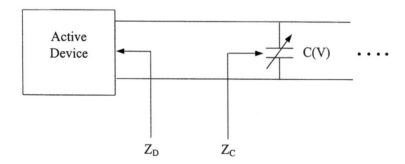

FIGURE 6.7 Varactor-tuned oscillator.

A high-efficiency transmitter is required for space or battery operating systems. The DC-to-RF conversion efficiency is given by

$$\eta = \frac{P_{RF}}{P_{DC}} \times 100\% \tag{6.4}$$

where P_{RF} is the generated RF power and P_{DC} is the DC bias power. In general, solid-state transistors or FETs can generate power ranging from milliwatts to a few watts with an efficiency ranging from 10 to 50%. Solid-state Gunn diodes can produce similar output power at a much lower efficiency of 1–3%. IMPATT diodes can produce several watts at 5–20% efficiency at high microwave or millimeter-wave frequencies.

For higher power, vacuum tubes such as traveling-wave tubes, Klystrons, or magnetrons can be used with efficiency ranging from 10 to 60%. Power-combining techniques can also be used to combine the power output from many low-power sources through chip-level, circuit-level, or spatial power combining [5].

In many cases, a high-power transmitter consists of a low-power oscillator followed by several stages of amplifiers. The first stage is called the driver amplifier, and the last stage is called the power amplifier. The power amplifier is normally one of the most expensive components in the system.

Example 6.1   A 35-GHz Gunn oscillator has a frequency variation of ±160 MHz over $-40°C$ to $+40°C$ temperature range. The oscillator can be tuned from 34.5 to 35.5 GHz with a varactor bias voltage varied from 0.5 to 4.5 V. What are the frequency stability in ppm/per degree Celsius and the frequency modulation sensitivity in megahertz per volts?

Solution

$$\text{Frequency stability} = \pm160 \text{ MHz}/80°C = \pm2 \text{ MHz}/°C$$

$$= A \text{ (in ppm}/°C) \times 10^{-6} \times 35 \times 10^{9} \text{ Hz}$$

$$A = \pm57 \text{ ppm}/°C$$

$$\text{Modulation sensitivity} = \frac{f_2 - f_1}{V_2 - V_1} = \frac{33.5 - 34.5 \text{ GHz}}{4.5 - 0.5 \text{ V}}$$

$$= 0.25 \text{ GHz/V} = 250 \text{ MHz/V} \qquad \blacksquare$$

6.5 INTERMODULATION

The intermodulation distortion and the third-order intercept point discussed in Chapter 5 for a receiver or mixer also apply to a power amplifier or upconverter in a transmitter. Figure 6.8 shows the curves for the fundamental and two-tone third-order intermodulation signals.

Conventional high-power RF/microwave amplifiers were once used to handle only a single carrier communication channel. In this case, they could operate within the nonlinear region of the dynamic range without the risk of intermodulation products generation, thus avoiding channel interference. Currently, many of the contemporary communication systems operate in a multicarrier environment that allows an enhancement in bandwidth efficiency. They are very attractive whenever there is a large demand to accommodate many users within a limited spectrum but are required to operate with minimized adjacent out-of-band spectral emissions (spectral containment). These unwanted frequency components are primarily the result of intermodulation distortion (IMD) products produced by the multiple carriers propagating through nonlinear solid-state devices.

Consider two signals f_1 and f_2 which are the input signals to a power amplifier, as shown in Fig. 6.9. The two signals will be amplified and the output power can be determined from the fundamental signal curve given in Fig. 6.8. The two-tone third-

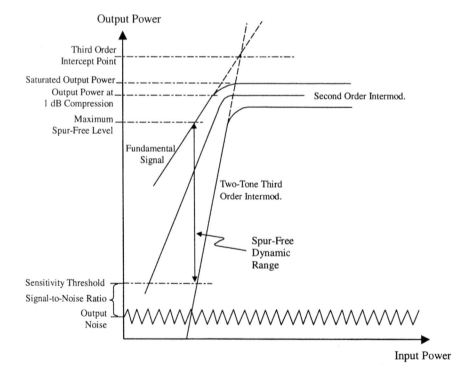

FIGURE 6.8 Nonlinear characteristics for a power amplifier.

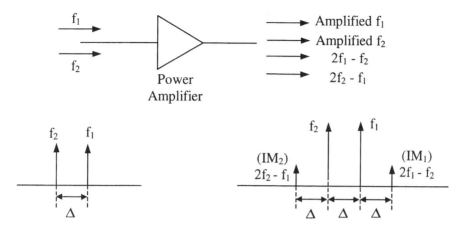

FIGURE 6.9   Power amplifier and its IM3 products.

order intermodulation products $(2f_1 - f_2$ and $2f_2 - f_1)$ are also generated and appear in the output port. The power levels of these IM products can be found from the two-tone third-order intermodulation (IM3) curve given in Fig. 6.8. The IM3 power levels are normally well below the fundamental signals at f_1 and f_2. If the frequency difference Δ is very small, the IM3 products are difficult to be filtered out, and it is important to keep their levels as low as possible. Other third-order distortion frequencies $3f_1$, $3f_2$, $2f_1 + f_2$, $2f_2 + f_1$, as well as the second-order distortion frequencies $2f_1$, $2f_2$, $f_1 + f_2$, $f_1 - f_2$, are of little concern because they are not closely adjacent in frequency and they can be easily filtered out without any disturbance to the original signals f_1 and f_2. In most wireless communications, one would like to have IM3 reduced to a level of less than -60 dBc (i.e., 60 dB or a million times below the fundamental signals).

One way to reduce the IM3 levels is to use the feedforward amplifier concept. The amplifier configuration consists of a signal cancellation loop and a distortion error cancellation loop, as shown in Fig. 6.10 [6]. The signal cancellation loop is composed of five elements: an equal-split power divider, a main power amplifier, a main-signal sampler, a phase/amplitude controller, and a power combiner. This loop samples part of the distorted signal out from the main amplifier and combines it with a previously adjusted, distortion-free sample of the main signal; consequently, the main signal is canceled and the IM products prevail. The error cancellation loop is composed of three elements: a phase/amplitude controller, a linear error amplifier, and an error coupler acting as a power combiner. This loop takes the IM products from the signal cancellation loop, adjusts their phase, increases their amplitude, and combines them with the signals from the main power amplifier in the error coupler. As a result, the third-order tones are greatly reduced to a level of less than -60 dBc. Experimental results are shown in Figs. 6.11 and 6.12.

Figure 6.11 shows the output of the main amplifier without the linearizer, where $f_1 = 2.165$ GHz and $f_2 = 2.155$ GHz. At these frequencies, $IM_1 = 2f_1 - f_2 =$

182

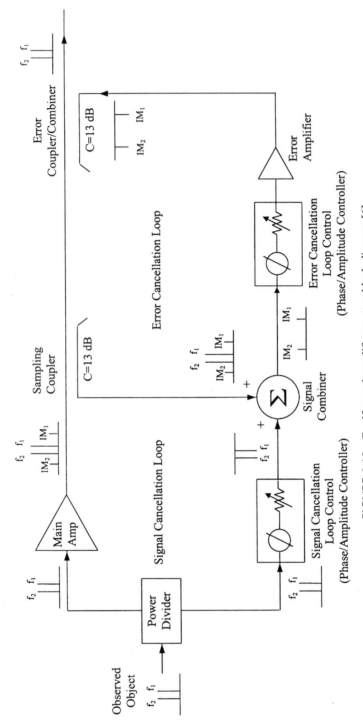

FIGURE 6.10 Feedforward amplifier system block diagram [6].

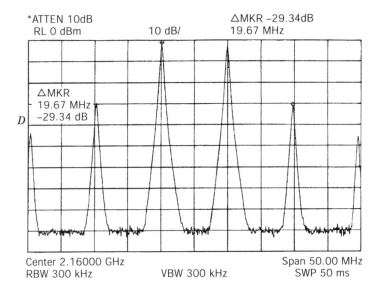

FIGURE 6.11 Nonlinearized two-tone test and intermodulation distortion [6].

2.175 GHz and $IM_2 = 2f_2 - f_1 = 2.145$ GHz, and the intermodulation distortion is approximately -30 dBc. Figure 6.12 shows the linearized two-tone test using the feedforward amplifier to achieve an additional 30 dB distortion reduction, giving a total IM suppression of -61 dBc

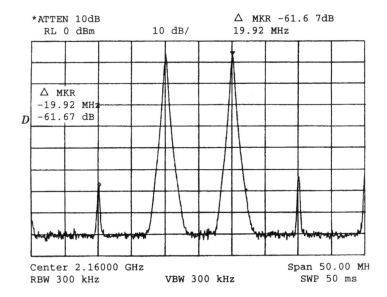

FIGURE 6.12 Linearized two-tone test and intermodulation distortion [6].

6.6   CRYSTAL REFERENCE OSCILLATORS

Crystal oscillators have low phase noise due to their stable output signal. The low-frequency crystal oscillators can be used as reference sources for a phase-locked loop. The crystal oscillator consists of a piezoelectric crystal, usually quartz, with both faces plated with electrodes. If a voltage is applied between the electrodes, mechanical forces will be exerted on the bound charges within the crystal, and an electromechanical system is formed that will vibrate at a resonant frequency. The resonant frequency and the Q factor depend on the crystal's dimensions and surface orientation. The Q's of several thousand to several hundred thousand and frequencies ranging from a few kilohertz to tens of megahertz are available. The extremely high Q values and the excellent stability of quartz with time and temperature give crystal oscillators the exceptional frequency stability.

The equivalent circuit of a crystal can be represented by Fig. 6.13. The inductor L, capacitor C, and resistor R represent the crystal. The capacitor C' represents the electrostatic capacitance between electrodes with the crystal as a dielectric. As an example, for a 90-kHz crystal, $L = 137$ H, $C = 0.0235$ pF, $C' = 3.5$ pF, with a Q of 5500. If we neglect R, the impedance of the crystal is a reactance shown in Fig. 6.14 given by

$$jX = -\frac{j}{\omega C'}\frac{\omega^2 - \omega_s^2}{\omega^2 - \omega_p^2} \tag{6.5}$$

where

$$\omega_s^2 = \frac{1}{LC} \quad \text{and} \quad \omega_p^2 = \frac{1}{L}\left(\frac{1}{C} + \frac{1}{C'}\right)$$

Since $C' \gg C$, $\omega_p^2 \approx 1/LC = \omega_s^2$. The circuit will oscillate at a frequency that lies between ω_s and ω_p. The oscillating frequency is essentially determined by the crystal and not by the rest of the circuit. The oscillating frequency is very stable

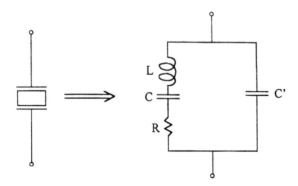

FIGURE 6.13   Piezoelectric crystal symbol and its equivalent circuit.

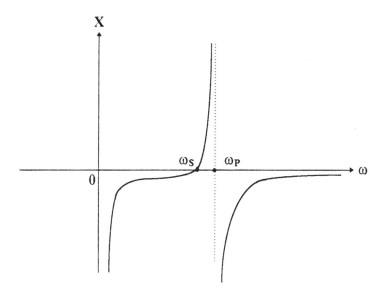

FIGURE 6.14 Impedance of a crystal as a function of frequency.

because $\omega_p \approx \omega_s$, where ω_s and ω_p are the series and parallel resonant frequencies. The crystal can be integrated into the transistor's oscillator circuit to form a crystal oscillator. Figure 6.15 shows two examples of these crystal oscillators [7]. In the next section, we will use the crystal oscillators to build high-frequency phase-locked oscillators (PLOs).

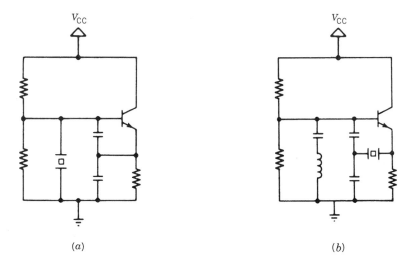

FIGURE 6.15 Colpitts crystal oscillators: (a) in parallel resonant configuration; (b) in series resonant configuration [7].

6.7 PHASE-LOCKED OSCILLATORS

A phase-locked oscillator is a very stable source with low phase noise and stable output frequency. A high-frequency oscillator can be phase locked to a low-frequency, stable crystal oscillator or crystal-controlled oscillator to achieve good phase noise and frequency stability. A simplified phase-locked loop (PLL) block diagram is given in Fig. 6.16. It consists of a very stable low-frequency oscillator that acts as the reference source, a phase detector, a low-pass filter, a VCO, and a frequency divider. The phase detector produces a DC control voltage at the output of the low-pass filter, with the magnitude and polarity determined by the phase (frequency) difference between the crystal oscillator and VCO output. The control voltage is used to vary the VCO frequency. The process will continue until the VCO frequency (or phase) is aligned with the multiple of the crystal oscillator frequency. The frequency divider is used to divide the output frequency of VCO by N to match the frequency of the reference oscillator. Because of the tracking, the output of the PLL has phase noise characteristics similar to that of the reference oscillator.

Figure 6.17 shows an example of an analog phase detector configuration that is similar to a balanced mixer. It consists of a 90° hybrid coupler and two mixer (detector) diodes, followed by a low-pass filter. If two signals of the nominally same frequency f_R but with different phases θ_1 and θ_2 are applied at the input of the coupler, the voltages across the mixer diodes are

$$v_1(t) = \cos(\omega_R t + \theta_1) + \cos(\omega_R t + \theta_2 - 90°)$$
$$= \cos(\omega_R t + \theta_1) + \sin(\omega_R t + \theta_2) \tag{6.6a}$$

$$v_2(t) = \cos(\omega_R t + \theta_2) + \cos(\omega_R t + \theta_1 - 90°)$$
$$= \cos(\omega_R t + \theta_2) + \sin(\omega_R t + \theta_1) \tag{6.6b}$$

Assuming that the diodes are operating in the square-law region, the output currents are given by

$$i_1(t) = A v_1^2(t)$$
$$= A[\cos^2(\omega_R t + \theta_1) + 2\cos(\omega_R t + \theta_1)\sin(\omega_R t + \theta_2) + \sin^2(\omega_R t + \theta_2)] \tag{6.7a}$$
$$i_2(t) = -A v_2^2(t)$$
$$= -A[\cos^2(\omega_R t + \theta_2) + 2\cos(\omega_R t + \theta_2)\sin(\omega_R t + \theta_1) + \sin^2(\omega_R t + \theta_1)] \tag{6.7b}$$

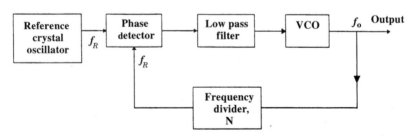

FIGURE 6.16 Simplified PLL block diagram.

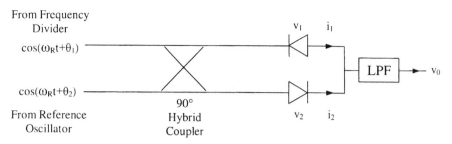

FIGURE 6.17   Analog phase detector.

where A is a constant. The negative sign on i_2 is due to the reversed diode polarity. The two currents will be combined and filtered through a low-pass filter. Now, the following trigonometry identities are used for Eqs. (6.7):

$$\cos^2 \alpha = \tfrac{1}{2}(1 + \cos 2\alpha) \qquad \sin^2 \alpha = \tfrac{1}{2}(1 - \cos 2\alpha)$$

and

$$2 \sin \alpha \cos \beta = \sin(\alpha + \beta) + \sin(\alpha - \beta)$$

Since the low-pass filter rejects all high-frequency components [i.e., $\cos 2\alpha$ and $\sin(\alpha + \beta)$ terms], only the DC current appears at the output

$$i_0(t) = i_1(t) + i_2(t) = A_1 \sin(\theta_2 - \theta_1) \tag{6.8}$$

where A_0 and A_1 are constants. Therefore, the output voltage of a phase detector is determined by the phase difference of its two input signals. For a small phase difference, we have

$$i_0(t) \approx A_1(\theta_2 - \theta_1) \tag{6.9}$$

The two input frequencies to the phase detector should be very close in order to be tracked (locked) to each other. The range of the input frequency for which the loop can acquire locking is called the capture range. The settling time is the time required for the loop to lock to a new frequency. Phase-locked loops can generate signals for FM, QPSK modulation, local oscillators for mixers, and frequency synthesizers. They are widely used in wireless communication systems.

Example 6.2   A 57-GHz phase-locked source has the block diagram shown in Fig. 6.18. Determine the reference signal frequency (f_R). The reference source is a crystal-controlled microwave oscillator. If the reference source has a frequency stability of ± 1 ppm/°C, what is the output frequency variation?

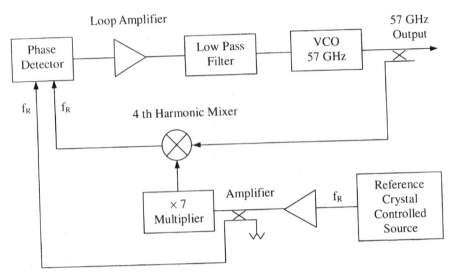

FIGURE 6.18 A 57-GHz phase-locked source. (From reference 8, with permission from IEEE.)

Solution In a harmonic mixer, the RF signal is mixed with the multiple frequency of the LO to generate an IF signal. The IF is given by

$$f_{IF} = f_{RF} - Nf_{LO}$$

or

$$f_{IF} = Nf_{LO} - f_{RF} \tag{6.10}$$

From Fig. 6.18, the reference frequency is given by

$$57 \text{ GHz} - 4 \times 7 \times f_R = f_R$$

Therefore, $f_R = 1.96551724$ GHz. The output frequency variation is

$$\Delta f = \pm f_0 \times 1 \text{ ppm/}°C = \pm 57 \times 10^9 \times 1 \times 10^{-6} \text{ Hz/}°C$$
$$= \pm 57 \text{ kHz/}°C$$

∎

6.8 FREQUENCY SYNTHESIZERS

A frequency synthesizer is a subsystem that derives a large number of discrete frequencies from an accurate, highly stable crystal oscillator. Each of the derived frequencies has the frequency stability and accuracy of the reference crystal source.

In many applications, the frequency synthesizer must cover a wide frequency range. A frequency synthesizer avoids the need for using many independent crystal-controlled oscillators in a wide-band multiple-channel system. Modern frequency synthesizers can be implemented using integrated circuit chips. They are controlled by digital circuits or computers. Frequency synthesizers are commonly used in transmitters, modulators, and LOs in many wireless communication systems such as radios, satellite receivers, cellular telephones, and data transmission equipment.

Frequency synthesizers can be realized using a PLL and a programmable frequency divider, as shown in Fig. 6.19. The signals applied to the phase detector are the reference signal from the crystal oscillator and f_0/N from the output of the frequency divider. A large number of frequencies can be obtained by varying N, the division ratio. As an example, if $f_R = 1$ MHz, we will have the output frequency (f_0) equal to 3 MHz, 4 MHz, ..., 20 MHz if $N = 3, 4, ..., 20$. The resolution or increment in frequency is equal to the reference frequency f_R. To improve the resolution, the reference frequency can also be divided before it is connected to the phase detector. This scheme is shown in Fig. 6.20. A fixed frequency divider with division ratio of N_2 is introduced between the crystal oscillator and the phase detector. Zero output from the phase detector requires the following condition:

$$\frac{f_0}{N_1} = \frac{f_R}{N_2} \tag{6.11}$$

Therefore

$$f_0 = \frac{N_1}{N_2}f_R = N_1\frac{f_R}{N_2} \tag{6.12}$$

The increment in frequency or resolution is equal to f_R/N_2. As an example,

$$f_R = 1 \text{ MHz} \qquad N_2 = 100 \qquad \text{Resolution} = 10 \text{ kHz}$$

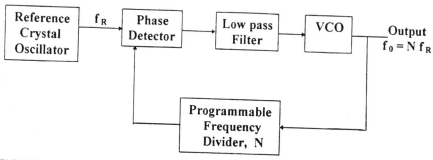

FIGURE 6.19   Frequency synthesizer using a PLL and a programmable frequency divider.

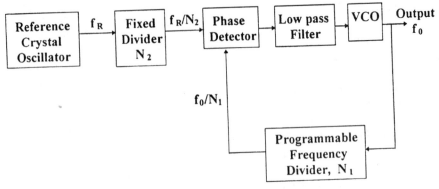

FIGURE 6.20   Frequency synthesizer with improved resolution.

If

$$N_1 = 3, \qquad f_0 = 30 \text{ kHz}$$
$$N_1 = 4 \qquad f_0 = 40 \text{ kHz}$$
$$\vdots \qquad \qquad \vdots$$

To obtain small frequency resolution and rapid frequency change, multiple-loop frequency synthesizers can be used; however, the system is more complicated. The following example shows a multiple-loop frequency synthesizer.

Example 6.3   In the multiple-loop frequency synthesizer shown in Fig. 6.21 [9], $f_R = 1$ MHz, $N_1 = 10$ and $N_2 = 100$. Determine the range of output frequencies of the synthesizer if N_A is varied from 200 to 300 and N_B from 350 to 400.

Solution

$$\frac{f_R N_A}{N_1 N_2} = f_0 - f_R \frac{N_B}{N_1}$$
$$f_0 = \frac{f_R}{N_1} \left(N_B + \frac{N_A}{N_2} \right)$$
$$f_R = 1 \text{ MHz} \qquad N_1 = 10 \qquad N_2 = 100$$

When $N_A = 200$, $N_B = 350$, we have $f_0 = f_{0(min)}$:

$$f_{0(min)} = \frac{1 \times 10^6}{10} \left(350 + \frac{200}{100} \right) = 35.2 \text{ MHz}$$

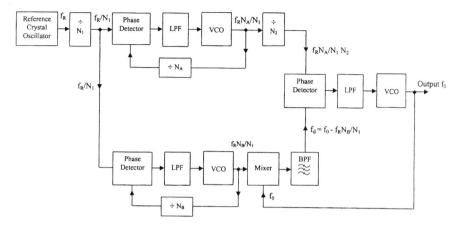

FIGURE 6.21   Multiple-loop frequency synthesizer [9].

When $N_A = 300$, $N_B = 400$, we have $f_0 = f_{0(max)}$:

$$f_{0(max)} = \frac{1 \times 10^6}{10}\left(400 + \frac{300}{100}\right) = 40.3 \text{ MHz}$$

The output range is from 35.2 to 40.3 MHz.   ■

PROBLEMS

6.1   A power amplifier has two input signals of $+10$ dBm at frequencies of 1.8 and 1.810 GHz (Fig. P6.1). The input IM3 power levels are -50 dBm. The amplifier has a gain of 10 dB and an input 1-dB compression point of $+25$ dBm. What are the frequencies for the IM3 products f_{IM1} and f_{IM2}? What are the power levels for f_{IM1} and f_{IM2} if the input power levels for f_1 and f_2 signals are increased to $+20$ dBm?

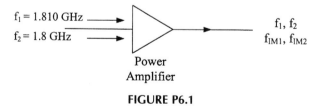

FIGURE P6.1

6.2   A power amplifier has two input signals at frequencies of 1 and 1.010 GHz. The output spectrums are shown in Fig. P6.2. What are the frequencies for the IM3 products f_{IM1} and f_{IM2}? If the output power levels for f_1 and f_2 signals are increased to 30 dBm, what are the power levels for f_{IM1} and f_{IM2} signals? Use $f_1 = 1.010$ GHz and $f_2 = 1$ GHz.

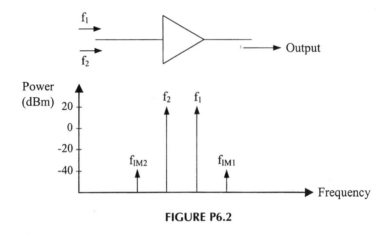

FIGURE P6.2

6.3   A 10-GHz PLO is shown in Fig. P6.3.

(a) Determine the reference frequency of the crystal-controlled source.

(b) If the reference source has a frequency stability of ± 0.1 ppm/°C, what is the output frequency variation of the PLO over the temperature range from -40 to $+40$°C?

(c) What is the reference frequency variation over the same temperature range?

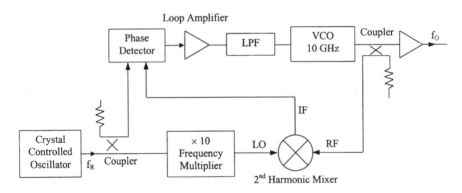

FIGURE P6.3

6.4   Calculate the reference signal frequency in gigahertz for the phase-locked system shown in Fig. P6.4.

6.5   Determine the output frequencies of the frequency synthesizer shown in Fig. P6.5 for $N_1 = 10$ and $N_1 = 20$. Note that $f_R = 1$ GHz and $N_2 = 100$.

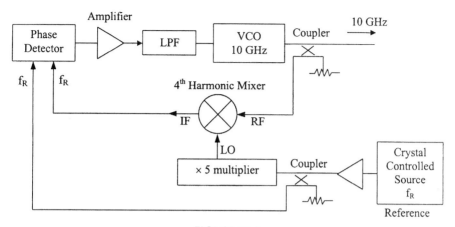

FIGURE P6.4

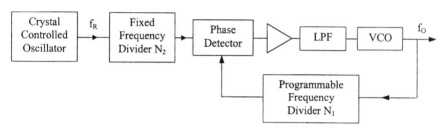

FIGURE P6.5

6.6   In the synthesizer shown in Fig. P6.6, $N_1 = 10$, $N_2 = 10$, and $f_R = 10$ MHz.

(a) What is the frequency resolution?

(b) If $N_3 = 1000$, what is the output frequency f_0? If $N_3 = 1001$, what is the output frequency f_0?

(c) If the frequency stability of the crystal reference oscillator is ± 1 ppm/°C, what is the frequency variation for the output signal over the temperature range from -30 to $+50$°C when $N_3 = 1000$?

6.7   A frequency synthesizer shown in Fig. P6.7 provides 401 output frequencies equally spaced by 10 kHz. The output frequencies are from 144 to 148 MHz.

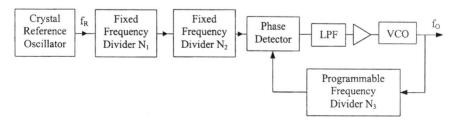

FIGURE P6.6

The reference frequency is 10 KHz, and the local oscillator frequency is 100 MHz. Calculate the minimum and maximum values for N.

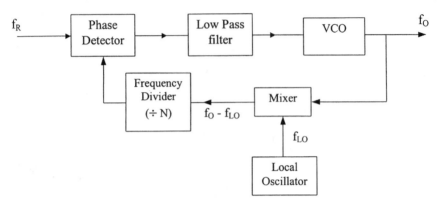

FIGURE P6.7

6.8   In Problem 6.7, if $N = 4600$, what is the output frequency?

6.9   In the synthesizer shown in Fig. P6.9, if $N_3 = 1000$ and $f_R = 1$ MHz, what is the output frequency for $N_1 = 100$ and $N_2 = 200$?

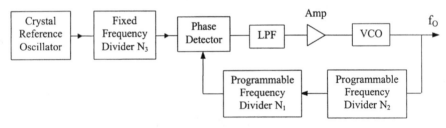

FIGURE P6.9

REFERENCES

1. *Watkin-Johnson Telecommunication Product Handbook*, Palo Alto, CA., 1996, p. 85.
2. G. D. Vendelin, A. M. Pavio, and U. L. Rhode, *Microwave Circuit Design*, John Wiley & Sons, New York, 1990, Ch. 6.
3. I. Bahl and P. Bhartia, *Microwave Solid State Circuit Design*, John Wiley & Sons, New York, 1988, Ch. 9.
4. A. L. Lance, "Microwave Measurements," in K. Chang, Ed., *Handbook of Microwave and Optical Components*, Vol. 1, John Wiley & Sons, New York, 1989, Ch. 9.
5. J. A. Navarro and K. Chang, *Integrated Active Antennas and Spatial Power Combining*, John Wiley & Sons, New York, 1996.
6. A. Echeverria, L. Fan, S. Kanamaluru, and K. Chang, "Frequency Tunable Feedforward Amplifier for PCS Applications," *Microwave Optic. Technol. Lett.*, Vol. 23, No. 4, pp. 218–221, 1999.

7. U. L. Rhode, *Microwave and Wireless Synthesizers*, John Wiley & Sons, New York, 1997.

8. K. Chang, K. Louie, A. J. Grote, R. S. Tahim, M. J. Mlinar, G. M. Hayashibara, and C. Sun, "V-Band Low-Noise Integrated Circuit Receiver," *IEEE Trans. Microwave Theory Tech.*, Vol. MTT-31, pp. 146–154, 1983.

9. D. C. Green, *Radio Systems Technology*, Longman Scientific & Technical, Essex, England, 1990.

CHAPTER SEVEN

Radar and Sensor Systems

7.1 INTRODUCTION AND CLASSIFICATIONS

Radar stands for radio detection and ranging. It operates by radiating electromagnetic waves and detecting the echo returned from the targets. The nature of an echo signal provides information about the target—range, direction, and velocity. Although radar cannot reorganize the color of the object and resolve the detailed features of the target like the human eye, it can see through darkness, fog and rain, and over a much longer range. It can also measure the range, direction, and velocity of the target.

A basic radar consists of a transmitter, a receiver, and a transmitting and receiving antenna. A very small portion of the transmitted energy is intercepted and reflected by the target. A part of the reflection is reradiated back to the radar (this is called back-reradiation), as shown in Fig. 7.1. The back-reradiation is received by the radar, amplified, and processed. The range to the target is found from the time it takes for the transmitted signal to travel to the target and back. The direction or angular position of the target is determined by the arrival angle of the returned signal. A directive antenna with a narrow beamwidth is generally used to find the direction. The relative motion of the target can be determined from the doppler shift in the carrier frequency of the returned signal.

Although the basic concept is fairly simple, the actual implementation of radar could be complicated in order to obtain the information in a complex environment. A sophisticated radar is required to search, detect, and track multiple targets in a hostile environment; to identify the target from land and sea clutter; and to discern the target from its size and shape. To search and track targets would require mechanical or electronic scanning of the antenna beam. For mechanical scanning, a motor or gimbal can be used, but the speed is slow. Phased arrays can be used for electronic scanning, which has the advantages of fast speed and a stationary antenna.

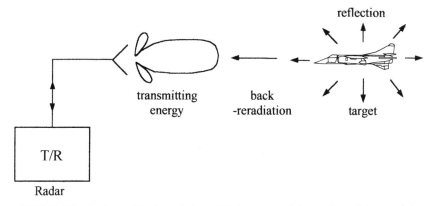

FIGURE 7.1 Radar and back-radiation: T/R is a transmitting and receiving module.

For some military radar, frequency agility is important to avoid lock-in or detection by the enemy.

Radar was originally developed during World War II for military use. Practical radar systems have been built ranging from megahertz to the optical region (laser radar, or ladar). Today, radar is still widely used by the military for surveillance and weapon control. However, increasing civil applications have been seen in the past 20 years for traffic control and navigation of aircraft, ships, and automobiles, security systems, remote sensing, weather forecasting, and industrial applications.

Radar normally operates at a narrow-band, narrow beamwidth (high-gain antenna) and medium to high transmitted power. Some radar systems are also known as sensors, for example, the intruder detection sensor/radar for home or office security. The transmitted power of this type of sensor is generally very low.

Radar can be classified according to locations of deployment, operating functions, applications, and waveforms.

1. *Locations*: airborne, ground-based, ship or marine, space-based, missile or smart weapon, etc.
2. *Functions*: search, track, search and track
3. *Applications*: traffic control, weather, terrain avoidance, collision avoidance, navigation, air defense, remote sensing, imaging or mapping, surveillance, reconnaissance, missile or weapon guidance, weapon fuses, distance measurement (e.g., altimeter), intruder detection, speed measurement (police radar), etc.
4. *Waveforms*: pulsed, pulse compression, continuous wave (CW), frequency-modulated continuous wave (FMCW)

Radar can also be classified as monostatic radar or bistatic radar. Monostatic radar uses a single antenna serving as a transmitting and receiving antenna. The transmitting and receiving signals are separated by a duplexer. Bistatic radar uses a separate transmitting and receiving antenna to improve the isolation between transmitter and receiver. Most radar systems are monostatic types.

Radar and sensor systems are big business. The two major applications of RF and microwave technology are communications and radar/sensor. In the following sections, an introduction and overview of radar systems are given.

7.2   RADAR EQUATION

The radar equation gives the range in terms of the characteristics of the transmitter, receiver, antenna, target, and environment [1, 2]. It is a basic equation for understanding radar operation. The equation has several different forms and will be derived in the following.

Consider a simple system configuration, as shown in Fig. 7.2. The radar consists of a transmitter, a receiver, and an antenna for transmitting and receiving. A duplexer is used to separate the transmitting and receiving signals. A circulator is shown in Fig. 7.2 as a duplexer. A switch can also be used, since transmitting and receiving are operating at different times. The target could be an aircraft, missile, satellite, ship, tank, car, person, mountain, iceberg, cloud, wind, raindrop, and so on. Different targets will have different radar cross sections (σ). The parameter P_t is the transmitted power and P_r is the received power. For a pulse radar, P_t is the peak pulse power. For a CW radar, it is the average power. Since the same antenna is used for transmitting and receiving, we have

$$G = G_t = G_r = \text{gain of antenna} \tag{7.1}$$

$$A_e = A_{et} = A_{er} = \text{effective area of antenna} \tag{7.2}$$

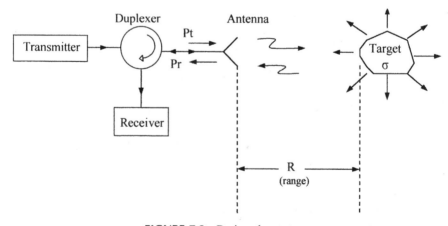

FIGURE 7.2   Basic radar system.

Note that

$$G_t = \frac{4\pi}{\lambda_0^2} A_{et} \tag{7.3}$$

$$A_{et} = \eta_a A_t \tag{7.4}$$

where λ_0 is the free-space wavelength, η_a is the antenna efficiency, and A_t is the antenna aperture size.

Let us first assume that there is no misalignment (which means the maximum of the antenna beam is aimed at the target), no polarization mismatch, no loss in the atmosphere, and no impedance mismatch at the antenna feed. Later, a loss term will be incorporated to account for the above losses. The target is assumed to be located in the far-field region of the antenna.

The power density (in watts per square meter) at the target location from an isotropic antenna is given by

$$\text{Power density} = \frac{P_t}{4\pi R^2} \tag{7.5}$$

For a radar using a directive antenna with a gain of G_t, the power density at the target location should be increased by G_t times. We have

$$\text{Power density at target location from a directive antenna} = \frac{P_t}{4\pi R^2} G_t \tag{7.6}$$

The measure of the amount of incident power intercepted by the target and reradiated back in the direction of the radar is denoted by the radar cross section σ, where σ is in square meters and is defined as

$$\sigma = \frac{\text{power backscattered to radar}}{\text{power density at target}} \tag{7.7}$$

Therefore, the backscattered power at the target location is [3]

$$\text{Power backscattered to radar (W)} = \frac{P_t G_t}{4\pi R^2} \sigma \tag{7.8}$$

A detailed description of the radar cross section is given in Section 7.4. The backscattered power decays at a rate of $1/4\pi R^2$ away from the target. The power density (in watts per square meters) of the echo signal back to the radar antenna location is

$$\text{Power density backscattered by target and returned to radar location} = \frac{P_t G_t}{4\pi R^2} \frac{\sigma}{4\pi R^2}$$

(7.9)

The radar receiving antenna captures only a small portion of this backscattered power. The captured receiving power is given by

$$P_r = \text{returned power captured by radar (W)} = \frac{P_t G_t}{4\pi R^2} \frac{\sigma}{4\pi R^2} A_{er}$$
(7.10)

Replacing A_{er} with $G_r \lambda_0^2/4\pi$, we have

$$P_r = \frac{P_t G_t}{4\pi R^2} \frac{\sigma}{4\pi R^2} \frac{G_r \lambda_0^2}{4\pi}$$
(7.11)

For monostatic radar, $G_r = G_t$, and Eq. (7.11) becomes

$$P_r = \frac{P_t G^2 \sigma \lambda_0^2}{(4\pi)^3 R^4}$$
(7.12)

This is the radar equation.

If the minimum allowable signal power is S_{min}, then we have the maximum allowable range when the received signal is $S_{i,min}$. Let $P_r = S_{i,min}$:

$$R = R_{max} = \left(\frac{P_t G^2 \sigma \lambda_0^2}{(4\pi)^3 S_{i,min}} \right)^{1/4}$$
(7.13)

where P_t = transmitting power (W)

G = antenna gain (linear ratio, unitless)

σ = radar cross section (m^2)

λ_0 = free-space wavelength (m)

$S_{i,min}$ = minimum receiving signal (W)

R_{max} = maximum range (m)

This is another form of the radar equation. The maximum radar range (R_{max}) is the distance beyond which the required signal is too small for the required system operation. The parameters $S_{i,\min}$ is the minimum input signal level to the radar receiver. The noise factor of a receiver is defined as

$$F = \frac{S_i/N_i}{S_o/N_o}$$

where S_i and N_i are input signal and noise levels, respectively, and S_o and N_o are output signal and noise levels, respectively, as shown in Fig. 7.3. Since $N_i = kTB$, as shown in Chapter 5, we have

$$S_i = kTBF\frac{S_o}{N_o} \tag{7.14}$$

where k is the Boltzmann factor, T is the absolute temperature, and B is the bandwidth. When $S_i = S_{i,\min}$, then $S_o/N_o = (S_o/N_o)_{\min}$. The minimum receiving signal is thus given by

$$S_{i,\min} = kTBF\left(\frac{S_o}{N_o}\right)_{\min} \tag{7.15}$$

Substituting this into Eq. (7.13) gives

$$R_{\max} = \left[\frac{P_t G^2 \sigma \lambda_0^2}{(4\pi)^3 kTBF\left(\dfrac{S_o}{N_o}\right)_{\min}}\right] \tag{7.16}$$

where $k = 1.38 \times 10^{-23}$ J/K, T is temperature in kelvin, B is bandwidth in hertz, F is the noise figure in ratio, $(S_o/N_o)_{\min}$ is minimum output signal-to-noise ratio in ratio. Here $(S_o/N_o)_{\min}$ is determined by the system performance requirements. For good probability of detection and low false-alarm rate, $(S_o/N_o)_{\min}$ needs to be high. Figure 7.4 shows the probability of detection and false-alarm rate as a function of (S_o/N_o). An S_o/N_o of 10 dB corresponds to a probability of detection of 76% and a false alarm probability of 0.1% (or 10^{-3}). An S_o/N_o of 16 dB will give a probability of detection of 99.99% and a false-alarm rate of 10^{-4}% (or 10^{-6}).

FIGURE 7.3   The SNR ratio of a receiver.

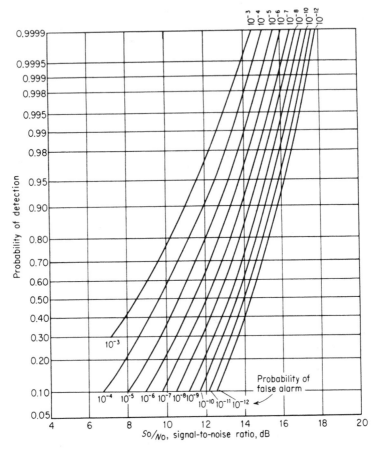

FIGURE 7.4   Probability of detection for a sine wave in noise as a function of the signal-to-noise (power) ratio and the probability of false alarm. (From reference [1], with permission from McGraw-Hill.)

7.3   RADAR EQUATION INCLUDING PULSE INTEGRATION AND SYSTEM LOSSES

The results given in Fig. 7.4 are for a single pulse only. However, many pulses are generally returned from a target on each radar scan. The integration of these pulses can be used to improve the detection and radar range. The number of pulses (n) on the target as the radar antenna scans through its beamwidth is

$$n = \frac{\theta_B}{\dot{\theta}_s} \times \text{PRF} = \frac{\theta_B}{\dot{\theta}_s} \frac{1}{T_p} \tag{7.17}$$

where θ_B is the radar antenna 3-dB beamwidth in degrees, $\dot{\theta}_s$ is the scan rate in degrees per second, PRF is the pulse repetition frequency in pulses per second, T_p is the period, and $\theta_B/\dot{\theta}_s$ gives the time that the target is within the 3-dB beamwidth of the radar antenna. At long distances, the target is assumed to be a point as shown in Fig. 7.5.

Example 7.1 A pulse radar system has a PRF = 300 Hz, an antenna with a 3-dB beamwidth of 1.5°, and an antenna scanning rate of 5 rpm. How many pulses will hit the target and return for integration?

Solution Use Eq. (7.17):

$$n = \frac{\theta_B}{\dot{\theta}_s} \times \text{PRF}$$

Now

$$\theta_B = 1.5° \qquad \dot{\theta}_s = 5 \text{ rpm} = 5 \times 360°/60 \text{ sec} = 30°/\text{sec}$$
$$\text{PRF} = 300 \text{ cycles/sec}$$
$$n = \frac{1.5°}{30°/\text{sec}} \times 300/\text{sec} = 15 \text{ pulses} \qquad ■$$

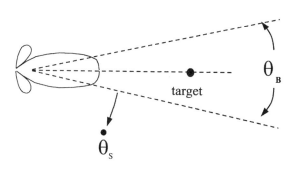

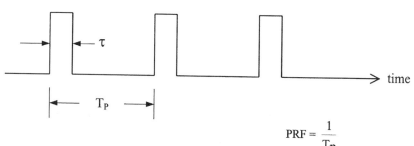

FIGURE 7.5 Concept for pulse integration.

Another system consideration is the losses involved due to pointing or misalignment, polarization mismatch, antenna feed or plumbing losses, antenna beam-shape loss, atmospheric loss, and so on [1]. These losses can be combined and represented by a total loss of L_{sys}. The radar equation [i.e., Eq. (7.16)] is modified to include the effects of system losses and pulse integration and becomes

$$R_{max} = \left[\frac{P_t G^2 \sigma \lambda_0^2 n}{(4\pi)^3 kTBF(S_o/N_o)_{min} L_{sys}} \right]^{1/4} \tag{7.18}$$

where $P_t =$ transmitting power, W

$G =$ antenna gain in ratio (unitless)

$\sigma =$ radar cross section of target, m^2

$\lambda_0 =$ free-space wavelength, m

$n =$ number of hits integrated (unitless)

$k = 1.38 \times 10^{-23}$ J/K (Boltzmann constant) (J $=$ W/sec)

$T =$ temperature, K

$B =$ bandwidth, Hz

$F =$ noise factor in ratio (unitless)

$(S_o/N_o)_{min} =$ minimum receiver output signal-to-noise ratio (unitless)

$L_{sys} =$ system loss in ratio (unitless)

$R_{max} =$ radar range, m

For any distance R, we have

$$R = \left[\frac{P_t G^2 \sigma \lambda_0^2 n}{(4\pi)^3 kTBF(S_o/N_o)L_{sys}} \right]^{1/4} \tag{7.19}$$

As expected, the S_o/N_o is increased as the distance is reduced.

Example 7.2   A 35-GHz pulse radar is used to detect and track space debris with a diameter of 1 cm [radar cross section (RCS) $= 4.45 \times 10^{-5}$ m^2]. Calculate the maximum range using the following parameters:

$$P_t = 2000 \text{ kW (peaks)} \qquad\qquad T = 290 \text{ K}$$
$$G = 66 \text{ dB} \qquad\qquad (S_o/N_o)_{min} = 10 \text{ dB}$$
$$B = 250 \text{ MHz} \qquad\qquad L_{sys} = 10 \text{ dB}$$
$$F = 5 \text{ dB} \qquad\qquad n = 10$$

Solution Substitute the following values into Eq. (7.18):

$$P_t = 2000 \text{ kW} = 2 \times 10^6 \text{ W} \qquad k = 1.38 \times 10^{-23} \text{ J/K}$$
$$G = 66 \text{ dB} = 3.98 \times 10^6 \qquad T = 290 \text{ K}$$
$$B = 250 \text{ MHz} = 2.5 \times 10^8 \text{ Hz} \qquad \sigma = 4.45 \times 10^{-5} \text{ m}^2$$
$$F = 5 \text{ dB} = 3.16 \qquad \lambda_0 = c/f_0 = 0.00857 \text{ m}$$
$$(S_o/N_o)_{\min} = 10 \text{ dB} = 10 \qquad L_{\text{sys}} = 10 \text{ dB} = 10$$
$$n = 10$$

Then we have

$$R_{\max} = \left[\frac{P_t G^2 \sigma \lambda_0^2 n}{(4\pi)^3 kTBF(S_o/N_o)_{\min} L_{\text{sys}}} \right]^{1/4}$$

$$= \left[\frac{2 \times 10^6 \text{ W} \times (3.98 \times 10^6)^2 \times 4.45 \times 10^{-5} \text{ m}^2 \times (0.00857 \text{ m})^2 \times 10}{(4\pi)^3 \times 1.38 \times 10^{-23} \text{ J/K} \times 290 \text{ K} \times 2.5 \times 10^8/\text{sec} \times 3.16 \times 10 \times 10} \right]^{1/4}$$

$$= 3.58 \times 10^4 \text{ m} = 35.8 \text{ km} \qquad \blacksquare$$

From Eq. (7.19), it is interesting to note that the strength of a target's echo is inversely proportional to the range to the fourth power $(1/R^4)$. Consequently, as a distant target approaches, its echoes rapidly grow strong. The range at which they become strong enough to be detected depends on a number of factors such as the transmitted power, size or gain of the antenna, reflection characteristics of the target, wavelength of radio waves, length of time the target is in the antenna beam during each search scan, number of search scans in which the target appears, noise figure and bandwidth of the receiver, system losses, and strength of background noise and clutter. To double the range would require an increase in transmitting power by 16 times, or an increase of antenna gain by 4 times, or the reduction of the receiver noise figure by 16 times.

7.4 RADAR CROSS SECTION

The RCS of a target is the effective (or fictional) area defined as the ratio of backscattered power to the incident power density. The larger the RCS, the higher the power backscattered to the radar.

The RCS depends on the actual size of the target, the shape of the target, the materials of the target, the frequency and polarization of the incident wave, and the incident and reflected angles relative to the target. The RCS can be considered as the effective area of the target. It does not necessarily have a simple relationship to the physical area, but the larger the target size, the larger the cross section is likely to be. The shape of the target is also important in determining the RCS. As an example, a corner reflector reflects most incident waves to the incoming direction, as shown in Fig. 7.6, but a stealth bomber will deflect the incident wave. The building material of

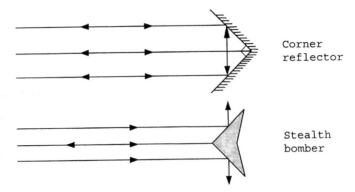

FIGURE 7.6   Incident and reflected waves.

the target is obviously an influence on the RCS. If the target is made of wood or plastics, the reflection is small. As a matter of fact, Howard Hughes tried to build a wooden aircraft (*Spruce Goose*) during World War II to avoid radar detection. For a metal body, one can coat the surface with absorbing materials (lossy dielectrics) to reduce the reflection. This is part of the reason that stealth fighters/bombers are invisible to radar.

The RCS is a strong function of frequency. In general, the higher the frequency, the larger the RCS. Table 7.1, comparing radar cross sections for a person [4] and various aircrafts, shows the necessity of using a higher frequency to detect small targets. The RCS also depends on the direction as viewed by the radar or the angles of the incident and reflected waves. Figure 7.7 shows the experimental RCS of a B-26 bomber as a function of the azimuth angle [5]. It can be seen that the RCS of an aircraft is difficult to specify accurately because of the dependence on the viewing angles. An average value is usually taken for use in computing the radar equation.

TABLE 7.1   Radar Cross Sections as a Function of Frequency

Frequency (GHz)			σ, m^2
(a) For a Person			
0.410			0.033–2.33
1.120			0.098–0.997
2.890			0.140–1.05
4.800			0.368–1.88
9.375			0.495–1.22
Aircraft	UHF	S-band, 2–4 GHz	X-band, 8–12 GHz
(b) For Aircraft			
Boeing 707	10 m^2	40 m^2	60 m^2
Boeing 747	15 m^2	60 m^2	100 m^2
Fighter	—	—	1 m^2

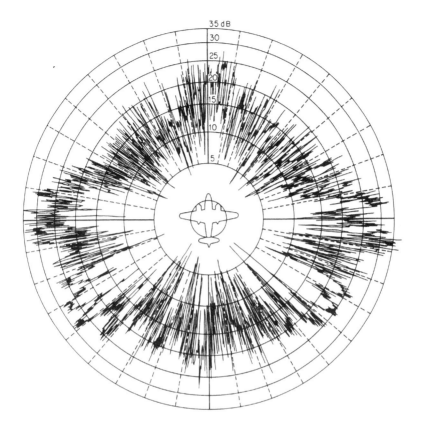

FIGURE 7.7   Experimental RCS of the B-26 bomber at 3 GHz as a function of azimuth angle [5].

For simple shapes of targets, the RCS can be calculated by solving Maxwell's equations meeting the proper boundary conditions. The determination of the RCS for more complicated targets would require the use of numerical methods or measurements. The RCS of a conducting sphere or a long thin rod can be calculated exactly. Figure 7.8 shows the RCS of a simple sphere as a function of its circumference measured in wavelength. It can be seen that at low frequency or when the sphere is small, the RCS varies as λ^{-4}. This is called the Rayleigh region, after Lord Rayleigh. From this figure, one can see that to observe a small raindrop would require high radar frequencies. For electrically large spheres (i.e., $a/\lambda \gg 1$), the RCS of the sphere is close to πa^2. This is the optical region where geometrical optics are valid. Between the optical region and the Rayleigh region is the Mie or resonance region. In this region, the RCS oscillates with frequency due to phase cancellation and the addition of various scattered field components.

Table 7.2 lists the approximate radar cross sections for various targets at microwave frequencies [1]. For accurate system design, more precise values

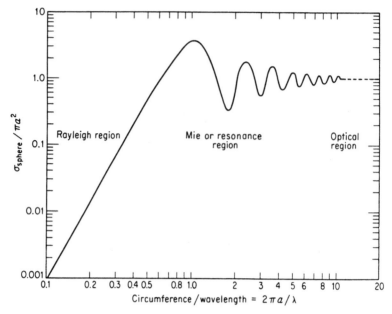

FIGURE 7.8 Radar cross section of the sphere: a = radius; λ = wavelength.

TABLE 7.2 Examples of Radar Cross Sections at Microwave Frequencies

	Cross Section (m³)
Conventional, unmanned winged missile	0.5
Small, single engine aircraft	1
Small fighter, or four-passenger jet	2
Large fighter	6
Medium bomber or medium jet airliner	20
Large bomber or large jet airliner	40
Jumbo jet	100
Small open boat	0.02
Small pleasure boat	2
Cabin cruiser	10
Pickup truck	200
Automobile	100
Bicycle	2
Man	1
Bird	0.01
Insect	10^{-5}

Source: From reference [1], with permission from McGraw-Hill.

should be obtained from measurements or numerical methods for radar range calculation. The RCS can also be expressed as dBSm, which is decibels relative to 1 m². An RCS of 10 m² is 10 dBSm, for example.

7.5   PULSE RADAR

A pulse radar transmits a train of rectangular pulses, each pulse consisting of a short burst of microwave signals, as shown in Fig. 7.9. The pulse has a width τ and a pulse repetition period $T_p = 1/f_p$, where f_p is the pulse repetition frequency (PRF) or pulse repetition rate. The duty cycle is defined as

$$\text{Duty cycle} = \frac{\tau}{T_p} \times 100\% \tag{7.20a}$$

The average power is related to the peak power by

$$P_{\text{av}} = \frac{P_t \tau}{T_p} \tag{7.20b}$$

where P_t is the peak pulse power.

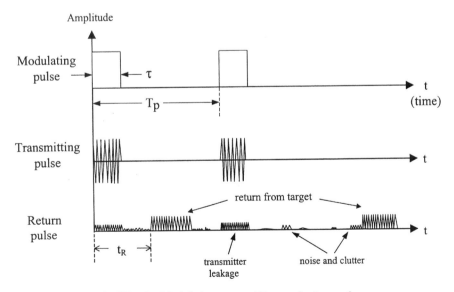

FIGURE 7.9   Modulating, transmitting, and return pulses.

The transmitting pulse hits the target and returns to the radar at some time t_R later depending on the distance, where t_R is the round-trip time of a pulsed microwave signal. The target range can be determined by

$$R = \tfrac{1}{2}ct_R \tag{7.21}$$

where c is the speed of light, $c = 3 \times 10^8$ m/sec in free space.

To avoid range ambiguities, the maximum t_R should be less than T_p. The maximum range without ambiguity requires

$$R'_{max} = \frac{cT_p}{2} = \frac{c}{2f_p} \tag{7.22}$$

Here, R'_{max} can be increased by increasing T_p or reducing f_p, where f_p is normally ranged from 100 to 100 kHz to avoid the range ambiguity.

A matched filter is normally designed to maximize the output peak signal to average noise power ratio. The ideal matched-filter receiver cannot always be exactly realized in practice but can be approximated with practical receiver circuits. For optimal performance, the pulse width is designed such that [1]

$$B\tau \approx 1 \tag{7.23}$$

where B is the bandwidth.

Example 7.3 A pulse radar transmits a train of pulses with $\tau = 10$ µs and $T_p = 1$ msec. Determine the PRF, duty cycle, and optimum bandwidth.

Solution The pulse repetition frequency is given as

$$\text{PRF} = \frac{1}{T_p} = \frac{1}{1 \text{ msec}} = 10^3 \text{ Hz}$$

$$\text{Duty cycle} = \frac{\tau}{T_p} \times 100\% = \frac{10 \text{ µsec}}{1 \text{ msec}} \times 100\% = 1\%$$

$$B = \frac{1}{\tau} = 0.1 \text{ MHz} \qquad \blacksquare$$

Figure 7.10 shows an example block diagram for a pulse radar system. A pulse modulator is used to control the output power of a high-power amplifier. The modulation can be accomplished either by bias to the active device or by an external $p-i-n$ or ferrite switch placed after the amplifier output port. A small part of the CW oscillator output is coupled to the mixer and serves as the LO to the mixer. The majority of output power from the oscillator is fed into an upconverter where it mixes with an IF signal f_{IF} to generate a signal of $f_0 + f_{IF}$. This signal is amplified by multiple-stage power amplifiers (solid-state devices or tubes) and passed through a duplexer to the antenna for transmission to free space. The duplexer could be a circulator or a transmit/receive (T/R) switch. The circulator diverts the signal moving from the power amplifier to the antenna. The receiving signal will be directed to the mixer. If it is a single-pole, double-throw (SPDT) T/R switch, it will be connected to the antenna and to the power amplifier in the transmitting mode and to the mixer in the receiving mode. The transmitting signal hits the target and returns to the radar antenna. The return signal will be delayed by t_R, which depends on the target range. The return signal frequency will be shifted by a doppler frequency (to be discussed in the next section) f_d if there is a relative speed between the radar and target. The return signal is mixed with f_0 to generate the IF signal of $f_{IF} \pm f_d$. The speed of the target can be determined from f_d. The IF signal is amplified, detected, and processed to obtain the range and speed. For a search radar, the display shows a polar plot of target range versus angle while the antenna beam is rotated for 360° azimuthal coverage.

To separate the transmitting and receiving ports, the duplexer should provide good isolation between the two ports. Otherwise, the leakage from the transmitter to the receiver is too high, which could drown the target return or damage the receiver. To protect the receiver, the mixer could be biased off during the transmitting mode, or a limiter could be added before the mixer. Another point worth mentioning is that the same oscillator is used for both the transmitter and receiver in this example. This greatly simplifies the system and avoids the frequency instability and drift problem. Any frequency drift in f_0 in the transmitting signal will be canceled out in the mixer. For short-pulse operation, the power amplifier can generate considerably higher

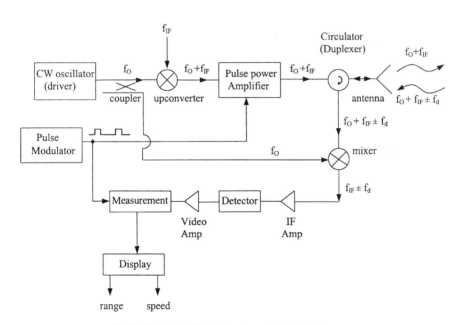

FIGURE 7.10 Typical pulse radar block diagram.

peak power than the CW amplifier. Using tubes, hundreds of kilowatts or megawatts of peak power are available. The power is much lower for solid-state devices in the range from tens of watts to kilowatts.

7.6   CONTINUOUS-WAVE OR DOPPLER RADAR

Continuous-wave or doppler radar is a simple type of radar. It can be used to detect a moving target and determine the velocity of the target. It is well known in acoustics and optics that if there is a relative movement between the source (oscillator) and the observer, an apparent shift in frequency will result. The phenomenon is called the doppler effect, and the frequency shift is the doppler shift. Doppler shift is the basis of CW or doppler radar.

Consider that a radar transmitter has a frequency f_0 and the relative target velocity is v_r. If R is the distance from the radar to the target, the total number of wavelengths contained in the two-way round trip between the target and radar is $2R/\lambda_0$. The total angular excursion or phase ϕ made by the electromagnetic wave during its transit to and from the target is

$$\phi = 2\pi \frac{2R}{\lambda_0} \tag{7.24}$$

The multiplication by 2π is from the fact that each wavelength corresponds to a 2π phase excursion. If the target is in relative motion with the radar, R and ϕ are continuously changing. The change in ϕ with respect to time gives a frequency shift ω_d. The doppler angular frequency shift ω_d is given by

$$\omega_d = 2\pi f_d = \frac{d\phi}{dt} = \frac{4\pi}{\lambda_0}\frac{dR}{dt} = \frac{4\pi}{\lambda_0}v_r \tag{7.25}$$

Therefore

$$f_d = \frac{2}{\lambda_0}v_r = \frac{2v_r}{c}f_0 \tag{7.26}$$

where f_0 is the transmitting signal frequency, c is the speed of light, and v_r is the relative velocity of the target. Since v_r is normally much smaller than c, f_d is very small unless f_0 is at a high (microwave) frequency. The received signal frequency is $f_0 \pm f_d$. The plus sign is for an approaching target and the minus sign for a receding target.

For a target that is not directly moving toward or away from a radar as shown in Fig. 7.11, the relative velocity v_r may be written as

$$v_r = v \cos \theta \tag{7.27}$$

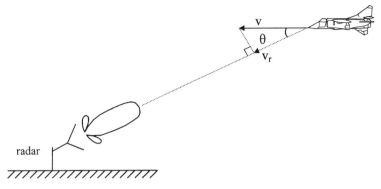

FIGURE 7.11   Relative speed calculation.

where v is the target speed and θ is the angle between the target trajectory and the line joining the target and radar. It can be seen that

$$v_r = \begin{cases} v & \text{if } \theta = 0 \\ 0 & \text{if } \theta = 90° \end{cases}$$

Therefore, the doppler shift is zero when the trajectory is perpendicular to the radar line of sight.

Example 7.4   A police radar operating at 10.5 GHz is used to track a car's speed. If a car is moving at a speed of 100 km/h and is directly aproaching the police radar, what is the doppler shift frequency in hertz?

Solution   Use the following parameters:

$$f_0 = 10.5 \text{ GHz}$$
$$\theta = 0°$$
$$v_r = v = 100 \text{ km/h} = 100 \times 1000 \text{ m/3600 sec} = 27.78 \text{ m/sec}$$

Using Eq. (7.26), we have

$$f_d = \frac{2v_r}{c} f_0 = \frac{2 \times 27.78 \text{ m/sec}}{3 \times 10^8 \text{ m/sec}} \times 10.5 \times 10^9 \text{ Hz}$$
$$= 1944 \text{ Hz} \qquad\blacksquare$$

Continuous-wave radar is relatively simple as compared to pulse radar, since no pulse modulation is needed. Figure 7.12 shows an example block diagram. A CW source/oscillator with a frequency f_0 is used as a transmitter. Similar to the pulse case, part of the CW oscillator power can be used as the LO for the mixer. Any frequency drift will be canceled out in the mixing action. The transmitting signal will

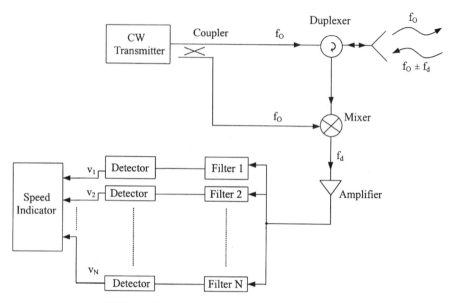

FIGURE 7.12   Doppler or CW radar block diagram.

pass through a duplexer (which is a circulator in Fig. 7.12) and be transmitted to free space by an antenna. The signal returned from the target has a frequency $f_0 \pm f_d$. This returned signal is mixed with the transmitting signal f_0 to generate an IF signal of f_d. The doppler shift frequency f_d is then amplified and filtered through the filter bank for frequency identification. The filter bank consists of many narrow-band filters that can be used to identify the frequency range of f_d and thus the range of target speed. The narrow-band nature of the filter also improves the SNR of the system. Figure 7.13 shows the frequency responses of these filters.

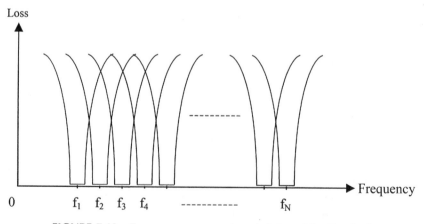

FIGURE 7.13   Frequency response characteristics of the filter bank.

Isolation between the transmitter and receiver for a single antenna system can be accomplished by using a circulator, hybrid junction, or separate polarization. If better isolation is required, separate antennas for transmitting and receiving can be used.

Since f_d is generally less than 1 MHz, the system suffers from the flicker noise ($1/f$ noise). To improve the sensitivity, an intermediate-frequency receiver system can be used. Figure 7.14 shows two different types of such a system. One uses a single antenna and the other uses two antennas.

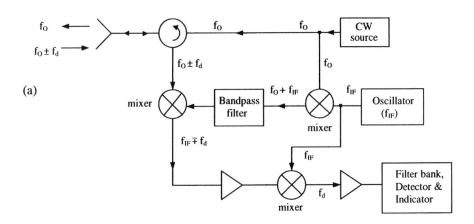

(a)

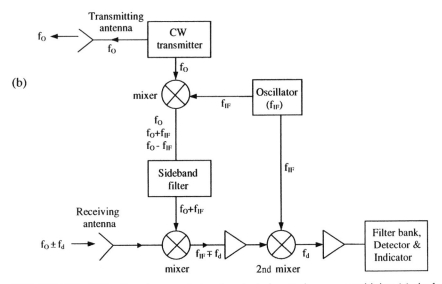

(b)

FIGURE 7.14 CW radar using superheterodyne technique to improve sensitivity: (*a*) single-antenna system; (*b*) two-antenna system.

The CW radar is simple and does not require modulation. It can be built at a low cost and has found many commercial applications for detecting moving targets and measuring their relative velocities. It has been used for police speed-monitoring radar, rate-of-climb meters for aircraft, traffic control, vehicle speedometers, vehicle brake sensors, flow meters, docking speed sensors for ships, and speed measurement for missiles, aircraft, and sports.

The output power of a CW radar is limited by the isolation that can be achieved between the transmitter and receiver. Unlike the pulse radar, the CW radar transmitter is on when the returned signal is received by the receiver. The transmitter signal noise leaked to the receiver limits the receiver sensitivity and the range performance. For these reasons, the CW radar is used only for short or moderate ranges. A two-antenna system can improve the transmitter-to-receiver isolation, but the system is more complicated.

Although the CW radar can be used to measure the target velocity, it does not provide any range information because there is no timing mark involved in the transmitted waveform. To overcome this problem, a frequency-modulated CW (FMCW) radar is described in the next section.

7.7 FREQUENCY-MODULATED CONTINUOUS-WAVE RADAR

The shortcomings of the simple CW radar led to the development of FMCW radar. For range measurement, some kind of timing information or timing mark is needed to recognize the time of transmission and the time of return. The CW radar transmits a single frequency signal and has a very narrow frequency spectrum. The timing mark would require some finite broader spectrum by the application of amplitude, frequency, or phase modulation.

A pulse radar uses an amplitude-modulated waveform for a timing mark. Frequency modulation is commonly used for CW radar for range measurement. The timing mark is the changing frequency. The transmitting time is determined from the difference in frequency between the transmitting signal and the returned signal.

Figure 7.15 shows a block diagram of an FMCW radar. A voltage-controlled oscillator is used to generate an FM signal. A two-antenna system is shown here for transmitter–receiver isolation improvement. The returned signal is $f_1 \pm f_d$. The plus sign stands for the target moving toward the radar and the minus sign for the target moving away from the radar. Let us consider the following two cases: The target is stationary, and the target is moving.

7.7.1 Stationary-Target Case

For simplicity, a stationary target is first considered. In this case, the doppler frequency shift (f_d) is equal to zero. The transmitter frequency is changed as a function of time in a known manner. There are many different forms of frequency–time variations. If the transmitter frequency varies linearly with time, as shown by the solid line in Fig. 7.16, a return signal (dotted line) will be received at t_R or $t_2 - t_1$ time later with $t_R = 2R/c$. At the time t_1, the transmitter radiates a signal with frequency f_1. When this signal is received at t_2, the transmitting frequency has been changed to f_2. The beat signal generated by the mixer by mixing f_2 and f_1 has a frequency of $f_2 - f_1$. Since the target is stationary, the beat signal (f_b) is due to the range only. We have

$$f_R = f_b = f_2 - f_1 \qquad (7.28)$$

From the small triangle shown in Fig. 7.16, the frequency variation rate is equal to the slope of the triangle:

$$\dot{f} = \frac{\Delta f}{\Delta t} = \frac{f_2 - f_1}{t_2 - t_1} = \frac{f_b}{t_R} \qquad (7.29)$$

The frequency variation rate can also be calculated from the modulation rate (frequency). As shown in Fig. 7.16, the frequency varies by $2\,\Delta f$ in a period of T_m, which is equal to $1/f_m$, where f_m is the modulating rate and T_m is the period. One can write

$$\dot{f} = \frac{2\,\Delta f}{T_m} = 2 f_m\,\Delta f \qquad (7.30)$$

Combining Eqs. (7.29) and (7.30) gives

$$f_b = f_R = t_R\,\dot{f} = 2 f_m t_R\,\Delta f \qquad (7.31)$$

Substituting $t_R = 2R/c$ into (7.31), we have

$$R = \frac{c f_R}{4 f_m\,\Delta f} \qquad (7.32)$$

The variation of frequency as a function of time is known, since it is set up by the system design. The modulation rate (f_m) and modulation range (Δf) are known. From Eq. (7.32), the range can be determined by measuring f_R, which is the IF beat frequency at the receiving time (i.e., t_2).

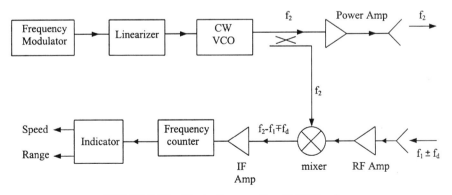

FIGURE 7.15 Block diagram of an FMCW radar.

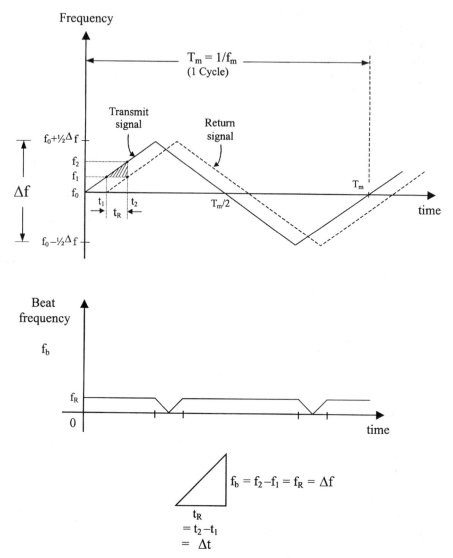

FIGURE 7.16 An FMCW radar with a triangular frequency modulation waveform for a stationary target case.

7.7.2 Moving-Target Case

If the target is moving, a doppler frequency shift will be superimposed on the range beat signal. It would be necessary to separate the doppler shift and the range information. In this case, f_d is not equal to zero, and the output frequency from the mixer is $f_2 - f_1 \mp f_d$, as shown in Fig. 7.15. The minus sign is for the target moving toward the radar, and the plus sign is for the target moving away from the radar.

Figure 7.17(*b*) shows the waveform for a target moving toward radar. For comparison, the waveform for a stationary target is also shown in Fig. 7.17(*a*). During the period when the frequency is increased, the beat frequency is

$$f_b \, (\text{up}) = f_R - f_d \tag{7.33}$$

During the period when the frequency is decreased, the beat frequency is

$$f_b \, (\text{down}) = f_R + f_d \tag{7.34}$$

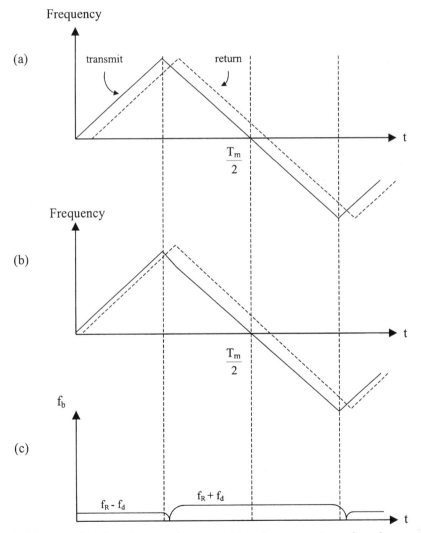

FIGURE 7.17 Waveform for a moving target: (*a*) stationary target waveform for comparison; (*b*) waveform for a target moving toward radar; (*c*) beat signal from a target moving toward radar.

The range information is in f_R, which can be obtained by

$$f_R = \tfrac{1}{2}[f_b \text{ (up)} + f_b \text{ (down)}] \tag{7.35}$$

The speed information is given by

$$f_d = \tfrac{1}{2}[f_b \text{ (down)} - f_b \text{ (up)}] \tag{7.36}$$

From f_R, one can find the range

$$R = \frac{cf_R}{4f_m \, \Delta f} \tag{7.37}$$

From f_d, one can find the relative speed

$$v_r = \frac{cf_d}{2f_0} \tag{7.38}$$

Similarly, for a target moving away from radar, one can find f_R and f_d from f_b (up) and f_b (down).

In this case, f_b (up) and f_b (down) are given by

$$f_b \text{ (up)} = f_R + f_d \tag{7.39}$$

$$f_b \text{ (down)} = f_R - f_d \tag{7.40}$$

Example 7.5   An FMCW altimeter uses a sideband superheterodyne receiver, as shown in Fig. 7.18. The transmitting frequency is modulated from 4.2 to 4.4 GHz linearly, as shown. The modulating frequency is 10 kHz. If a returned beat signal of 20 MHz is detected, what is the range in meters?

Solution   Assuming that the radar is pointing directly to the ground with $\theta = 90°$, we have

$$v_r = v \cos \theta = 0$$

From the waveform, $f_m = 10$ kHz and $\Delta f = 200$ MHz.

The beat signal $f_b = 20$ MHz $= f_R$. The range can be calculated from Eq. (7.32):

$$R = \frac{cf_R}{4f_m \, \Delta f} = \frac{3 \times 10^8 \text{ m/sec} \times 20 \times 10^6 \text{ Hz}}{4 \times 10 \times 10^3 \text{ Hz} \times 200 \times 10^6 \text{ Hz}}$$

$$= 750 \text{ m}$$

Note that both the range and doppler shift can be obtained if the radar antenna is tilted with $\theta \neq 90°$.   ∎

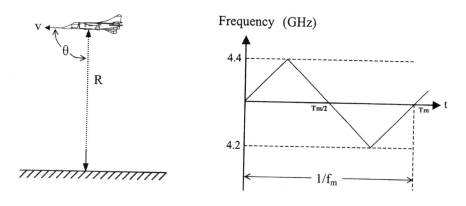

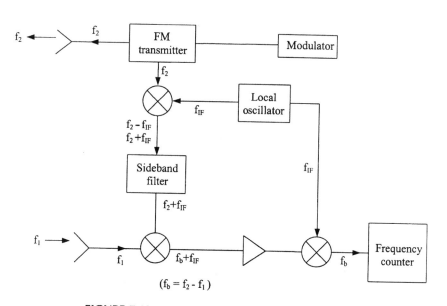

FIGURE 7.18 An FMCW altimeter and its waveform.

For an FMCW radar with a linear waveform, as shown in Fig. 7.16, a perfect linear curve is assumed in the calculation of the range. Any deviation from the linear curve will affect the accuracy and resolution in the range calculation. A linearizer can be used between the voltage-controlled oscillator and the varactor bias supply to produce a linear frequency tuning.

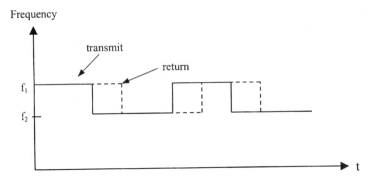

FIGURE 7.19 Two-step frequency modulation waveform.

Other frequency waveforms can also be used. Figure 7.19 shows a two-step frequency modulation as a function of time. The beat signal will occur when the frequency of the returning signal differs from the transmitting signal.

7.8 DIRECTION FINDING AND TRACKING

A tracking radar needs to find its target before it can track. The radar first searches for the target. This is normally done by scanning the antenna beam over a wide angle in one- or two-dimensional space. The beam scanning can be accomplished mechanically or electronically or both. After the target is detected, the tracking radar will track the target by measuring the coordinates of a target and providing data that may be used to determine the target path and to predict its future position. To predict future positions, all or part of the following data are required: range, speed, elevation angle, and azimuth angle. The elevation and azimuth angles give the direction of the target. A radar might track the target in range, in angle, in doppler shift, or any combination of the three. But in general, a tracking radar refers to angle tracking.

It is difficult to optimize a radar to operate in both search and tracking modes. Therefore, many radar systems employ a separate search radar and tracking radar. The search radar or acquisition radar provides the target coordinates for the tracking radar. The tracking radar acquires the target by performing a limited search in the area of the designated target coordinates provided by the search radar.

There are two major types of tracking radar: continuous-tracking radar and track-while-scan radar. The continuous-tracking radar provides continuous-tracking data on a particular target, while the track-while-scan radar supplies sampled data on one or more targets.

The ability of a radar to find the direction of a target is due to a directive antenna with a narrow beamwidth. An error signal will be generated if the antenna beam is not exactly on the target. This error signal will be used to control a servomotor to align the antenna to the target. There are three methods for generating the error signal for tracking:

1. Sequential lobing
2. Conical scan
3. Monopulse

A brief description of each method is given below.

7.8.1 Sequential Lobing

In sequential lobing, the antenna beam is switched between two positions, as shown in Fig. 7.20. The difference in amplitude between the received signal voltages obtained in the two switched positions is a measure of the angular displacement of the target from the switching axis.

The sign of the difference determines the direction that the antenna must be moved in order to align the switching axis with the direction of the target. When the

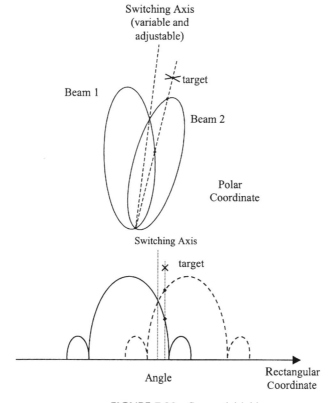

FIGURE 7.20 Sequential lobing.

voltages in the two switched positions are equal, the target is tracked in the direction of the switching axis.

7.8.2 Conical Scan

As shown in Fig. 7.21, this method is an extension of sequential lobing. The antenna rotates continuously instead of switching to two discrete positions. If the rotation axis is not aligned with the target, the echo signal will be amplitude modulated at a frequency equal to the rotation frequency of the beam. The demodulated signal is applied to the servocontrol system to continuously adjust the antenna position. When the line of sight to the target and rotation axis are aligned, there is no amplitude modulation output (i.e., returning signal has the same amplitude during rotation).

7.8.3 Monopulse Tracking

The above two methods require a minimum number of pulses to extract the angle error signal. In the time interval during which a measurement is made, the train of echo pulses must contain no amplitude modulation components other than the modulation produced by scanning. The above two methods could be severely limited in applications due to the fluctuating target RCS. Monopulse tracking uses a *single* pulse rather than many pulses and avoids the effects of pulse-to-pulse amplitude fluctuation.

For simplicity, let us consider the one-dimensional monopulse tracking first. Example applications are surface-to-surface or ship-to-ship tracking radar. The amplitude comparison monopulse radar uses two overlapping antenna patterns to obtain the azimuth angular error in one coordinate, as shown in Fig. 7.22. The sum

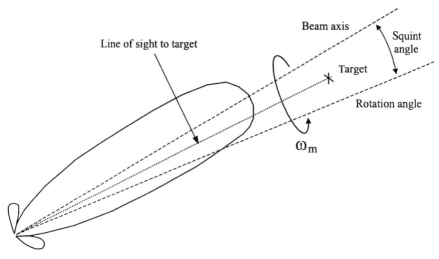

FIGURE 7.21 Conical scan.

and difference of the antenna patterns and error signal are shown in Fig. 7.23. If the target is aligned with the antenna axis, $A = B$ and the angle error signal is zero. The target is tracked. Otherwise, $A \neq B$, resulting in an error signal. The one-dimensional monopulse system consists of two antenna feeds located side by side, a transmitter, two receivers (one for the sum channel and the other for the difference channel), and a monopulse comparator. A block diagram is shown in Fig. 7.24. The comparator generates the sum-and-difference signals. Microwave hybrid couplers, magic-Ts, or rat-race ring circuits can be used as comparators. Shown in Fig. 7.24 is a rat-race ring circuit. The transmitting signal at port 1 will split equally into ports 2 and 3 in phase and radiate out through the two feeds. The received signals at ports 2 and 3 will arrive at port 1 in phase, resulting in a sum ($\sum$) signal, and arrive at port 4 with $180°$ out of phase, resulting in a difference (Δ) signal. The different phases are due to the different signal traveling paths, as shown in Fig. 7.24. The antenna beams are used for direction finding and tracking. They can also be used in homing and navigation applications.

Mathematically, one can derive $A + B$ and $A - B$ if the target is located at an angle θ from the broad side, as shown for Fig. 7.25. The phase difference between the two paths is

$$\phi = \frac{2\pi}{\lambda_0} d \sin \theta = 2\pi \frac{d \sin \theta}{\lambda_0} \tag{7.41}$$

where d is the separation of the two antenna feeds and λ_0 is the free-space wavelength:

$$|A + B| = |V_0 e^{j\phi} + V_0| = |V_0(1 + e^{j\phi})|$$

$$= |V_0|\sqrt{(1 + \cos \phi)^2 + \sin^2 \phi}$$

$$= 2|V_0| \cos(\tfrac{1}{2} \phi) \tag{7.42}$$

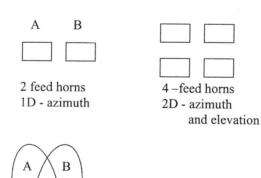

A   B

2 feed horns
1D - azimuth

4 –feed horns
2D - azimuth
    and elevation

A   B

Antenna Pattern

FIGURE 7.22   Feed arrangements of monopulse systems.

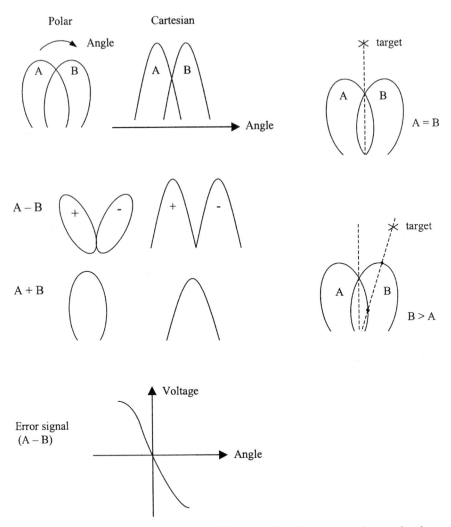

FIGURE 7.23   The sum $(A + B)$ and differences $(A - B)$ patterns and error signal.

When $\theta = 0$, the target is tracked. We have $\phi = 0$ and $|A + B| = 2|V_0| = $ maximum.

$$|A - B| = |V_0 - V_0 e^{j\phi}| = |V_0|\sqrt{(1 - \cos \phi)^2 + \sin^2 \phi}$$
$$= 2|V_0| \sin \tfrac{1}{2} \phi \qquad (7.43)$$

When $\theta = 0$, we have $\phi = 0$ and $|A - B| = 0$. As shown in Fig. 7.25, $|A + B|$ and $|A - B|$ can be plotted as a function of ϕ.

If both the azimuth and elevation tracking are needed, a two-dimensional monopulse system is required. Figure 7.26 shows a block diagram for the

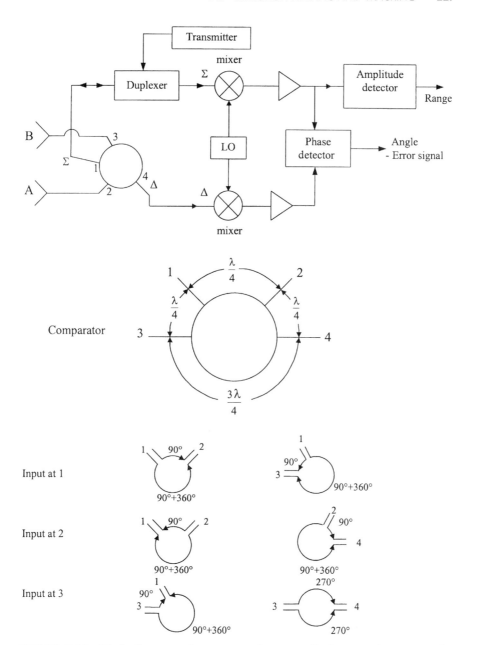

FIGURE 7.24 Block diagram and comparator for an amplitude comparison monopulse radar with one angular coordinate.

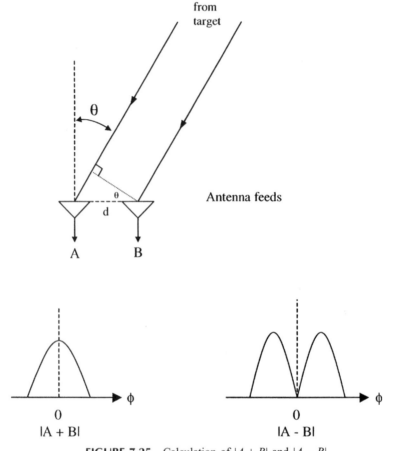

FIGURE 7.25   Calculation of $|A + B|$ and $|A - B|$.

comparator and Fig. 7.27 for the receiver. Three separate receivers are used to process the sum $(A + B + C + D)$, the elevation error $[(B + D) - (A + C)]$, and the azimuth error $[(A + B) - (C + D)]$ signals. The sum signal is used as a reference for the phase detector and also to extract the range information. The signal $[(A + D) - (B + C)]$ is not useful and dumped to a load.

7.9   MOVING-TARGET INDICATION AND PULSE DOPPLER RADAR

The doppler shift in the returning pulse produced by a moving target can be used by a pulse radar to determine the relative velocity of the target. This application is similar to the CW radar described earlier. The doppler shift can also be used to separate the moving target from undesired stationary fixed background (clutter). This application allows the detection of a small moving target in the presence of large

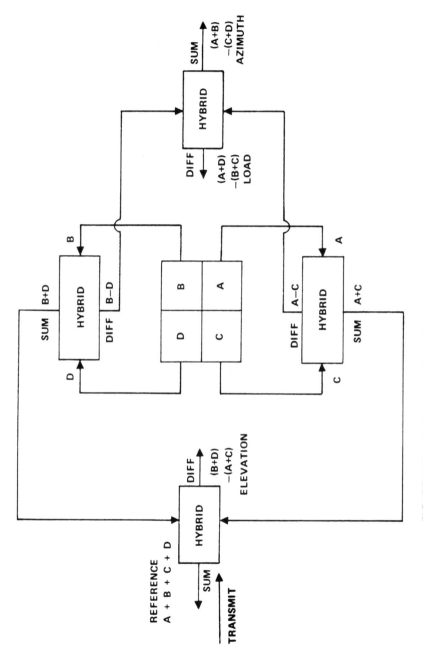

FIGURE 7.26 Configuration of amplitude comparison monopulse system.

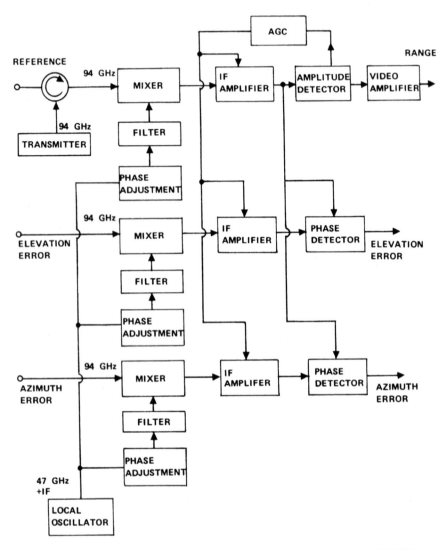

FIGURE 7.27 Amplitude comparison monopulse receiver operating at 94 GHz.

clutter. Such a pulse radar that utilizes the doppler frequency shift as a means for discriminating the moving target from the fixed clutter is called an MTI (moving target indicator), or pulse doppler radar. For example, a small moving car in the presence of land clutter and a moving ship in sea clutter can be detected by an MTI radar.

Figure 7.28 shows a simple block diagram for an MTI radar. The transmitting pulse has a carrier frequency f_0. The returning pulse frequency is $f_0 \pm f_d$. The

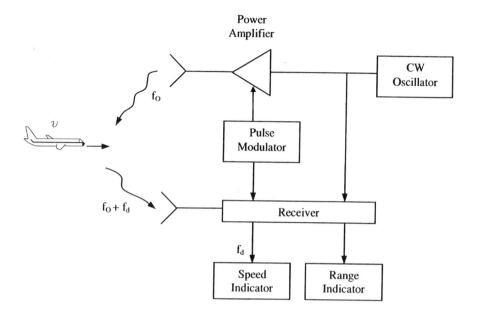

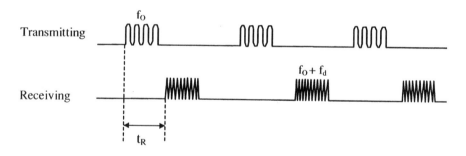

FIGURE 7.28 Pulse doppler radar.

returning pulse is also delayed by t_R, which is the time for the signal to travel to the target and return. The range and relative velocity can be found by

$$R = \tfrac{1}{2}ct_R \qquad (7.44)$$

$$v_r = \frac{cf_d}{2f_0} \qquad (7.45)$$

The delay line cancelers, range-gated doppler filters, and digital signal processing are used to extract the doppler information [1].

7.10   SYNTHETIC APERTURE RADAR

A synthetic aperture radar (SAR) achieves high resolution in the cross-range dimension by taking advantage of the motion of the aircraft or satellite carrying the radar to synthesize the effect of a large antenna aperture. The use of SAR can generate high-resolution maps for military reconnaissance, geological and mineral exploration, and remote sensing applications [1, 6].

Figure 7.29 shows an aircraft traveling at a constant speed v along a straight line. The radar emits a pulse every pulse repetition period (T_p). During this time, the airplane travels a distance vT_p. The crosses in the figure indicate the position of the radar antenna when a pulse is transmitted. This is equivalent to a linear synthesized antenna array with the element separation of a distance vT_p.

Conventional aperture resolution on the cross-range dimension is given by [1]

$$\delta = R\theta_B \tag{7.46}$$

where R is the range and θ_B is the half-power beamwidth of the antenna. For an SAR, the resolution becomes

$$\delta_s = R\theta_s = \tfrac{1}{2}D \tag{7.47}$$

where θ_s is the beamwidth of a synthetic aperture antenna and D is the diameter of the antenna. It is interesting to note that δ_s is independent of the range.

As an example, for a conventional antenna with $\theta_B = 0.2°$ (this is a very narrow beamwidth) at a range of 100 km, we have, from Eq. (7.46),

$$\delta = 350 \text{ m}$$

For an SAR at X-band with an antenna diameter of 3 m, the resolution is

$$\delta_s = 1.5 \text{ m}$$

which is much better than δ.

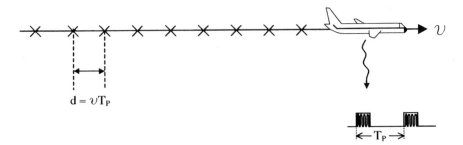

FIGURE 7.29   Synthetic aperture radar traveling at a constant speed v.

7.11   PRACTICAL RADAR EXAMPLES

There are many considerations in radar system design. The user requirements have a direct impact on the microwave subsystem design as summarized in Table 7.3 [7].

Table 7.4 shows the system parameters for a commercial air route surveillance radar. The L-band frequency was selected for all-weather operations with a range of

TABLE 7.3   Impact of User Requirements on Microwave Design

User Requirement	Impact on Microwave Subsystems
Type of target (scattering cross section) and maximum/minimum range	Antenna size and operating frequency; transmitter power; transmitter types: solid state or tube; waveform
Ability to observe close-in targets	Sensitivity time control (STC) provided by $p-i-n$ diode attenuator in RF or IF
Coverage	Antenna rotation (azimuth and/or elevation); mechanically versus electronically scanned antenna
Number of targets observed simultaneously	Mechanically versus electronically scanned antenna
Minimum spacing of targets	Antenna size; operating frequency (higher frequencies yield more spatial resolution); pulse width (modulation frequency)
Operating weather conditions	Operating frequency; increase transmitter power to overcome attenuation due to weather
Site restrictions	Transmitter and antenna size; transmitter efficiency (affects size of power conditioning circuitry); blank sector (turn off transmitter at specified azimuth angles); reduce ground clutter (use phase stable transmitter, higher operating frequency)
Reliability/availability	Redundant transmitters, power supplies, receivers; switchover box
Maintainability	Replacable subsystems (consider interface specification tolerances to eliminate retuning circuits); connectors positioned for ease of operation
Electromagnetic compatibility	Receiver has high dynamic range to minimize spurious signals; transmitter pulse shaping to restrict bandwidth; transmitter filtering to avoid spurious signals
Weight	Efficient transmitter reduces supply weight; lightweight antenna affects performance
Cost/delivery schedule	Use known, proven designs; introduce new technologies where needed; begin development/testing of "untried" components early

Source: From reference [7].

TABLE 7.4   Parameters of ARSR-3 Radar System[a]

Minimum target size	2 m^2
Maximum range	370 km (200 nautical miles)
Operating frequency range	1.25–1.35 GHz
Probability of detection	80% (one scan)
Probability of false alarm	10^{-6}
Pulse width	2 sec
Resolution	500 m
Pulse repetition frequency	310–365 Hz
Lowest blind speed	600 m/sec (1200 knots)
Peak power	5 MW
Average power	3600 W
Receiver noise figure	4 dB
Moving-target indicator (MTI) improvement	39 dB (3-pulse canceler, 50 dB with 4-pulse canceler)
Antenna size	12.8 m (42 ft) × 6.9 m (22 ft)
Beamwidth (azimuth/elevation)	1.25/40°
Polarization	Horizontal, vertical, and circular
Antenna gain	34 dB (34.5 dB lower beam, 33.5 dB upper beam)
Antenna scan time	12 sec
EIRP (on-axis effective isotropic radiate power)	101 dBW (12,500 MW)
Average power/effective aperture/scan time product (antenna efficiency = 60%)	2300 kW m^2 sec

[a]Courtesy of Westinghouse Electric Corp.
Source: From reference [7].

370 km, a probability of detection of 80% for one scan, and a false-alarm rate of 10^{-6} (i.e., one false report per a million pulses) for a target size of 2 m^2. The high peak power is generated by a Klystron tube. Figure 7.30 shows the radar installation. The antenna is a truncated reflector. A radome is used to protect the antenna. Dual systems operate simultaneously on opposite polarizations at slightly different frequencies. This allows operation with one system shut down for repairs.

Table 7.5 shows the system parameters and performance of a USSR Cosmos 1500 side-looking radar [2]. The oceanographic satellite was launched on September 28, 1983, into a nominal 650-km polar orbit to provide continuous world ocean observations for civil and military users. The radar serves as a radiometer to monitor oceans and ice zones, using a slotted waveguide array antenna. The system operates at 9.5 GHz with a peak output power of 100 kW generated by a magnetron tube.

Space-based radars in satellites can be used to monitor global air traffic. A rosette constellation of three satellites at an orbital altitude of 10,371 km provides continuous worldwide visibility by at least two satellites simultaneously [8]. Table 7.6 shows the system parameters. The radar antenna is a phased array with a maximum scan angle of 22.4°. Distributed solid-state T/R modules are used as transmitters. Each module provides 155 mW peak power with a total of 144,020 modules.

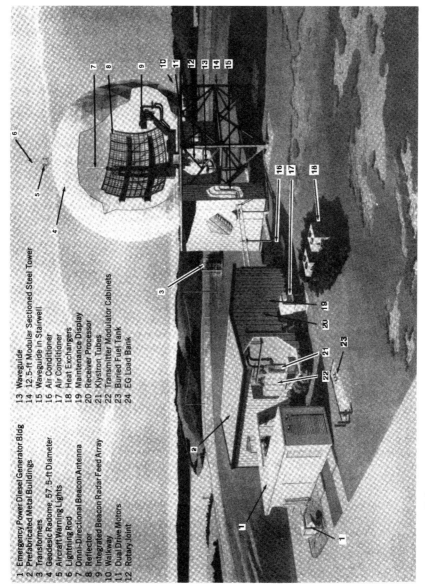

1 Emergency Power Diesel Generator Bldg
2 Prefabricated Metal Buildings
3 Transformers
4 Geodesic Radome, 57.5-ft Diameter
5 Aircraft Warning Lights
6 Lightning Rod
7 Omni-Directional Beacon Antenna
8 Reflector
9 Integrated Beacon Radar Feed Array
10 Walkway
11 Dual Drive Motors
12 Rotary Joint

13 Waveguide
14 12.5-ft Modular Sectioned Steel Tower
15 Waveguide in Stairwell
16 Air Conditioner
17 Air Conditioner
18 Heat Exchangers
19 Maintenance Display
20 Receiver Processor
21 Klystron Tubes
22 Transmitter Modulator Cabinets
23 Buried Fuel Tank
24 EG Load Bank

FIGURE 7.30 ARSR-3 radar station diagram. (Courtesy of Westinghouse Electric Corp.)

235

TABLE 7.5 Cosmos 1500 SLR Parameters and Performance

Type	Real-beam side-looking radar (460-km swath)
Frequency/wavelength	9500 MHz/3.15 cm
Antenna	
Type	Slotted waveguide
Size	11.085 m × 40 mm
No. of slots	480
Illumination	Cosine of a pedestal
Beamwidth	0.20° × 42°
Gain	35 dB
Sidelobes	−22 dB to −25 dB
Waveguide	Copper, 23 × 10-mm cross section
Polarization	Vertical
Swing angle	35° from nadir
Noise temperature	300 K
Transmitter	
Type	Magnetron
Power	100 kW peak, 30 W average
Pulse width	3 μsec
PRF	100 pps
Loss	1.7 dB
Receiver	
Type	Superheterodyne
Noise power	−140 dBW
Loss	1.7 dB
Pulses integrated	8 noncoherent
LNA noise temperature	150–200 K
LNA gain	15 dB
Dynamic range	30 dB
IF	30 ± 0.1 MHz
Input power	400 W
Range	700 km (minimum), 986 km (maximum)
SNR	0 dB on $\sigma^0 = -20$ dB

Source: From reference [2].

PROBLEMS

7.1 A radar transmits a peak power of 20 kW at 10 GHz. The radar uses a dish antenna with a gain of 40 dB. The receiver has a bandwidth of 25 MHz and a noise figure of 10 dB operating at room temperature (290 K). The overall system loss is estimated to be 6 dB. The radar is used to detect an airplane with a radar cross section of 10 m^2 at a distance of 1 km. (a) What is the return signal level (P_r) (in watts)? (b) What is the SNR (in decibels) at the receiver output? (Assume only one pulse integration.)

TABLE 7.6   Radar Parameters for Global Air Traffic Surveillance

Antenna	
Type	Corporate-fed active phased array
Diameter	100 m
Frequency	2 GHz
Wavelength	0.15 m
Polarization	Circular
Number of elements	576,078
Number of modules	144,020
Element spacing	0.7244 wavelength
Beamwidth	1.83 mrad
Directive gain	66.42 dB
Maximum scan angle	22.4°
Receiver	
Type	Distributed solid-state monolithic T/R module
Bandwidth	500 kHz
System noise temperature	490 K
Compressed pulse width	2 μsec
Transmitter	
Type	Distributed solidd-state monolithic T/R module
Peak power	22.33 kW
Pulse width	200 μsec
Maximum duty	0.20
Frequency	2 GHz
Signal Processor	
Type	Digital
Input speed	50 million words per second

Source: From reference [8], with permission from IEEE.

7.2 A high-power radar uses a dish antenna with a diameter of 4 m operating at 3 GHz. The transmitter produces 100 kW CW power. What is the minimum safe distance for someone accidentally getting into the main beam? (The U.S. standard requires a power density of < 10 mW/cm^2 for safety.) Assume that the dish antenna has a 55% efficiency.

7.3 A pulse radar has the following parameters:

$$P_t = 16 \text{ kW (peak)} \quad \text{Frequency} = 35 \text{ GHz}, \quad T = 290 \text{ K}$$
$$B = 20 \text{ MHz} \quad \text{PRF} = 100 \text{ kHz}, \quad F = 6 \text{ dB}$$
$$\text{Pulse with } \tau = 0.05 \text{ μsec} \quad \text{Scan rate} = 60°/\text{sec} \quad L_{\text{sys}} = 10 \text{ dB}$$

$$\text{Antenna} = 2\text{-m dish (i.e., diameter} = 2 \text{ m)}$$

$$\text{Target (airplane) RCS} = 30 \text{ m}^2 \quad \text{Antenna efficiency} = 55\%$$

$$\text{Probability of detection} = 0.998 \quad \text{False-alarm rate} = 10^{-8}$$

Calculate (a) the maximum range in kilometers, (b) the receiver output SNR when the target is 140 km from the radar, and (c) the maximum range in kilometers if the peak power is increased to 160 kW.

7.4   A pulse radar has the following parameters: $P_t = 100$ kW, PRF $= 10$ kHz, pulse width $= 2$ µsec. What are the duty cycle in percent, the average transmit power in watts, and the maximum range (in meters) without ambiguity?

7.5   A pulse radar has the following parameters:

$$P_t = 1000 \text{ W (peak)} \quad \text{Frequency} = 35 \text{ GHz}, \qquad\qquad T = 290 \text{ K}$$
$$\text{PRF} = 100 \text{ kHz} \qquad\qquad\qquad B = 20 \text{ MHz}, \quad \text{Pulse width } \tau = 0.05 \text{ µsec}$$
$$F = 6 \text{ dB} \qquad\qquad \text{Scan rate} = 60°/\text{sec} \qquad\qquad L_{\text{sys}} = 10 \text{ dB}$$

$$\text{Antenna} = 2\text{-m dish (i.e., diameter} = 2 \text{ m)}$$

$$\text{Airplane (target) RCS} = 30 \text{ m}^2 \qquad \text{Antenna efficiency} = 55\%$$

$$\text{Probability of detection} = 0.998 \qquad \text{False-alarm rate} = 10^{-8}$$

Calculate (a) the maximum range in kilometers and (b) the maximum range without ambiguity in kilometers. (c) Determine which one is the limiting maximum range.

7.6   A pulse radar is used for air traffic control. The radar uses a 55% efficiency dish antenna with a diameter of 3 m operating at 10 GHz. The airplane has a radar cross section of 60 m² and is located 180 km away from the radar. Assume that $P_t = 100$ kW, $T = 290$ K, $B = 10$ MHz, $F = 6$ dB, $L_{\text{sys}} = 10$ dB, PRF $= 1$ kHz, and $n = 10$. Calculate (a) the SNR at the output of the radar receiver, (b) the SNR at the input of the radar receiver, (c) the probability of detection if the false-alarm rate is 10^{-12}, and (d) the scan rate in degrees per second.

7.7   A pulse radar is used for surveillance. The radar is operating at 20 GHz with a dish antenna. The dish antenna has a diameter of 2 m and an efficiency of 55%. An airplane has a radar cross section of 50 m² located at a distance of 500 km away from the radar. Other system parameters are $P_t = 100$ kW, temperature $= 290$ K, $B = 20$ MHz, $F = 6$ dB, $L_{\text{sys}} = 10$ dB, PRF $= 1$ kHz, and $n = 10$. Calculate (a) the SNR at the output of the radar receiver, (b) the SNR at the input of the radar receiver, (c) the false-alarm rate if the probability of detection is 95%, and (d) the scan rate in degrees per second.

7.8   A pulse radar has the following specifications:

$$\text{Pulse width} = 1 \text{ µsec}$$
$$\text{Pulse repetition frequency} = 1 \text{ kHz}$$
$$\text{Peak power} = 100 \text{ kW}$$

Calculate (a) the duty cycle in percent, (b) the maximum range in kilometers without ambiguity, and (c) the average power in watts.

7.9 A radar is used for air defense surveillance to detect a fighter with a radar cross section of 1 m². The radar transmits a peak power of 100 kW at 10 GHz. The antenna gain is 50 dB. The system has a bandwidth of 10 MHz and operates at room temperature. The receiver has a noise figure of 6 dB. The overall system loss is 6 dB. If we want to have a probability of detection of 90% and a false-alarm rate of 0.01% (i.e., 10^{-4}), what is the maximum range of this radar assuming only one pulse integration?

7.10 A pulsed radar for air defense uses a dish antenna with a diameter of 6 m operating at 3 GHz. The output peak power is 100 kW. The radar repetition rate is 1 kHz and the pulse width is 1 μsec. The scan rate of the antenna beam is 3 rpm. Calculate (a) the duty cycle in percent, (b) the average transmit power in watts, (c) the time (in seconds) that the target is within the 3-dB beamwidth of the radar antenna, (d) the number of pulses returned to radar, (e) the maximum range without ambiguity in meters, and (f) the optimum system bandwidth.

7.11 A car travels at a speed of 120 km/h. A police radar is positioned at the location shown in Fig. P7.11. The radar is operating at 24.5 GHz. Calculate the doppler shift frequency that appears as the IF frequency at the output of the radar mixer.

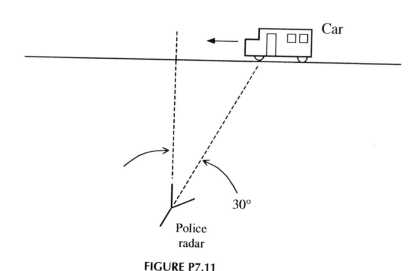

FIGURE P7.11

7.12 A CW radar is used to track an airplane, as shown in Fig. P7.12. If the radar transmits a 10-GHz signal and the airplane travels at a speed of 800 km/h, what is the returned signal frequency?

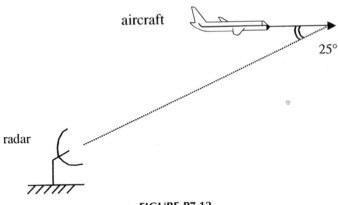

FIGURE P7.12

7.13   A CW radar is used to track the speed of an aircraft, as shown in Fig. P7.13. If the aircraft travels at a speed of 1000 km/h and the radar transmits a signal of 10 GHz, what is the returned signal frequency?

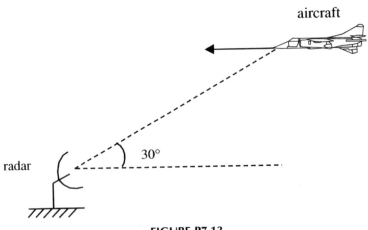

FIGURE P7.13

7.14   The transmitting and receiving signal frequencies as a function of time of an FMCW radar are shown in Fig. P7.14. The target is stationary. Determine the range (in kilometers) of the target if the output IF beat signal is (a) 25 MHz and (b) 10 MHz.

7.15   An FMCW radar has the returned beat signal (f_b) as a function of time, as shown in Fig. P7.15. Calculate the range and the speed of the target.

7.16   An FMCW radar has the return beat signal frequency shown in Fig. P7.16. The operating center frequency is 20 GHz and the frequency is varied in the range of 20 GHz ± 10 MHz. Calculate the range in kilometers and the relative velocity in kilometers per second for the target.

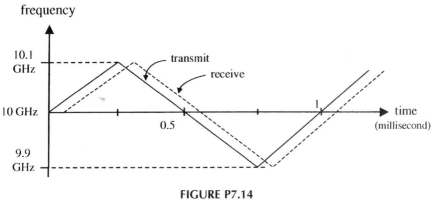

FIGURE P7.14

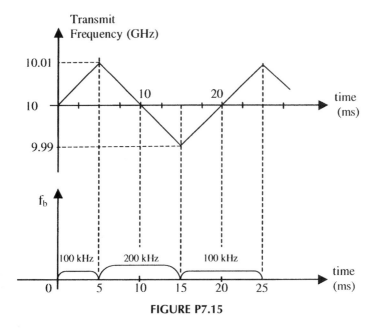

FIGURE P7.15

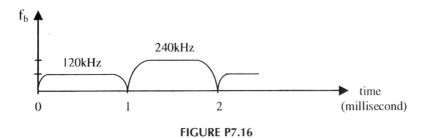

FIGURE P7.16

REFERENCES

1. M. L. Skolnik, *Introduction to Radar Systems*, 2nd ed., McGraw-Hill, New York, 1980.
2. M. L. Skolnik, Ed., *Radar Handbook*, McGraw Hill, 2nd ed., New York, 1990.
3. D. M. Pozar, *Microwave Engineering*, 2nd ed., John Wiley & Sons, New York, 1998.
4. F. V. Schultz, R. C. Burgener, and S. King, "Measurement of the Radar Cross Section of a Man," *Proc. IRE*, Vol. 46, pp. 476–481, 1958.
5. L. N. Ridenour, "Radar System Engineering," *MIT Radiation Laboratory Series*, Vol. I, McGraw-Hill, New York, 1947.
6. J. C. Curlander and R. N. McDonough, *Synthetic Aperture Radar*, John Wiley & Sons, New York, 1991.
7. E. A. Wolff and R. Kaul, *Microwave Engineering and Systems Applications*, John Wiley & Sons, New York, 1988.
8. L. J. Cantafio and J. S. Avrin, "Satellite-Borne Radar for Global Air Traffic Surveillance," IEEE ELECTRO '82 Convention, Boston, May 25–27, 1982.

CHAPTER EIGHT

Wireless Communication Systems

8.1 INTRODUCTION

The RF and microwave wireless communication systems include radiolinks, tropo-scatter/diffraction, satellite systems, cellular/cordless/personal communication systems (PCSs)/personal communication networks (PCNs), and wireless local-area networks (WLANs). The microwave line-of-sight (LOS) point-to-point radiolinks were widely used during and after World War II. The LOS means the signals travel in a straight line. The LOS link (or hop) typically covers a range up to 40 miles. About 100 LOS links can cover the whole United States and provide transcontinental broadband communication service. The troposcatter (scattering and diffraction from troposphere) can extend the microwave LOS link to several hundred miles. After the late 1960s, geostationary satellites played an important role in telecommunications by extending the range dramatically. A satellite can link two points on earth separated by 8000 miles (about a third of the way around the earth). Three such satellites can provide services covering all major population centers in the world. The satellite uses a broadband system that can simultaneously support thousands of telephone channels, hundreds of TV channels, and many data links. After the mid-1980s, cellular and cordless phones became popular. Wireless personal and cellular communications have enjoyed the fastest growth rate in the telecommunications industry. Many satellite systems are being deployed for wireless personal voice and data communications from any part of the earth to another using a hand-held telephone or laptop computer.

243

8.2 FRIIS TRANSMISSION EQUATION

Consider the simplified wireless communication system shown in Fig. 8.1. A transmitter with an output power P_t is fed into a transmitting antenna with a gain G_t. The signal is picked up by a receiving antenna with a gain G_r. The received power is P_r and the distance is R. The received power can be calculated in the following if we assume that there is no atmospheric loss, polarization mismatch, impedance mismatch at the antenna feeds, misalignment, and obstructions. The antennas are operating in the far-field regions.

The power density at the receiving antenna for an isotropic transmitting antenna is given as

$$S_I = \frac{P_t}{4\pi R^2} \quad (\text{W/m}^2) \tag{8.1}$$

Since a directive antenna is used, the power density is modified and given by

$$S_D = \frac{P_t}{4\pi R^2} G_t \quad (\text{W/m}^2) \tag{8.2}$$

The received power is equal to the power density multiplied by the effective area of the receiving antenna

$$P_r = \frac{P_t G_t}{4\pi R^2} A_{er} \quad (\text{W}) \tag{8.3}$$

The effective area is related to the antenna gain by the following expression:

$$G_r = \frac{4\pi}{\lambda_0^2} A_{er} \quad \text{or} \quad A_{er} = \frac{G_r \lambda_0^2}{4\pi} \tag{8.4}$$

Substituting (8.4) into (8.3) gives

$$P_r = P_t \frac{G_t G_r \lambda_0^2}{(4\pi R)^2} \tag{8.5}$$

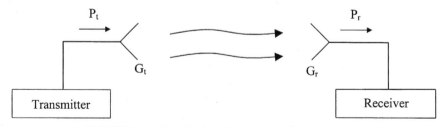

P_t P_r

G_t G_r

Transmitter   Receiver

FIGURE 8.1 Simplified wireless communication system.

This equation is known as the Friis power transmission equation. The received power is proportional to the gain of either antenna and inversely proportional to R^2.

If $P_r = S_{i,min}$, the minimum signal required for the system, we have the maximum range given by

$$R_{max} = \left[\frac{P_t G_t G_r \lambda_0^2}{(4\pi)^2 S_{i,min}} \right]^{1/2}$$
(8.6)

To include the effects of various losses due to misalignment, polarization mismatch, impedance mismatch, and atmospheric loss, one can add a factor L_{sys} that combines all losses. Equation (8.6) becomes

$$R_{max} = \left[\frac{P_t G_t G_r \lambda_0^2}{(4\pi)^2 S_{i,min} L_{sys}} \right]^{1/2}$$
(8.7)

where $S_{i,min}$ can be related to the receiver parameters. From Fig. 8.2, it can be seen that the noise factor is defined in Chapter 5 as

$$F = \frac{S_i/N_i}{S_o/N_o}$$
(8.8)

Therefore

$$S_i = S_{i,min} = N_i F \left(\frac{S_o}{N_o} \right)_{min}$$

$$= kTBF \left(\frac{S_o}{N_o} \right)_{min}$$
(8.9)

where k is the Boltzmann constant, T is the absolute temperature, and B is the receiver bandwidth. Substituting (8.9) into (8.7) gives

$$R_{max} = \left[\frac{P_t G_t G_r \lambda_0^2}{(4\pi)^2 kTBF(S_o/N_o)_{min} L_{sys}} \right]^{1/2}$$
(8.10)

S_i S_o

Receiver

N_i N_o

S_i / N_i S_o / N_o

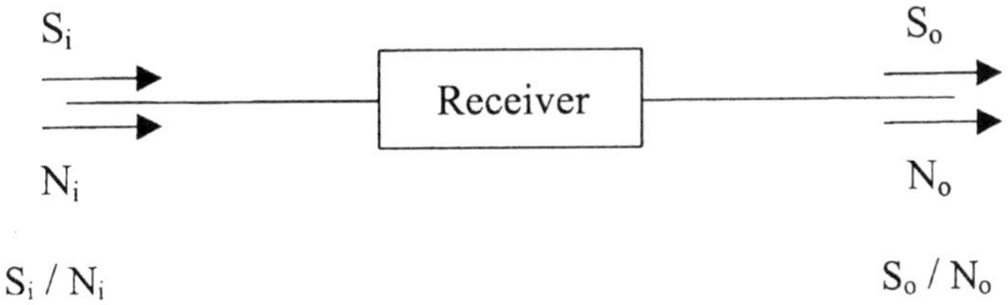

FIGURE 8.2 Receiver input and output SNRs.

where P_t = transmitting power (W)

G_t = transmitting antenna gain in ratio (unitless)

G_r = receiving antenna gain in ratio (unitless)

λ_0 = free-space wavelength (m)

$k = 1.38 \times 10^{-23}$ J/K (Boltzmann constant)

T = temperature (K)

B = bandwidth (Hz)

F = noise factor (unitless)

$(S_o/N_o)_{min}$ = minimum receiver output SNR (unitless)

L_{sys} = system loss in ratio (unitless)

R_{max} = maximum range (m)

The output SNR for a distance of R is given as

$$\frac{S_o}{N_o} = \frac{P_t G_t G_r}{kTBFL_{sys}} \left(\frac{\lambda_0}{4\pi R}\right)^2 \tag{8.11}$$

From Eq. (8.10), it can be seen that the range is doubled if the output power is increased four times. In the radar system, it would require the output power be increased by 16 times to double the operating distance.

From (Eq. 8.11), it can be seen that the receiver output SNR ratio can be increased if the transmission distance is reduced. The increase in transmitting power or antenna gain will also enhance the output SNR ratio as expected.

Example 8.1 In a two-way communication, the transmitter transmits an output power of 100 W at 10 GHz. The transmitting antenna has a gain of 36 dB, and the receiving antenna has a gain of 30 dB. What is the received power level at a distance of 40 km (a) if there is no system loss and (b) if the system loss is 10 dB?

Solution

$$f = 10 \text{ GHz} \qquad \lambda_0 = \frac{c}{f} = 3 \text{ cm} = 0.03 \text{ m}$$

$$P_t = 100 \text{ W} \qquad G_t = 36 \text{ dB} = 4000 \qquad G_r = 30 \text{ dB} = 1000$$

(a) From Eq. (8.5),

$$P_r = P_t \frac{G_t G_r \lambda_0^2}{(4\pi R)^2}$$

$$= 100 \times \frac{4000 \times 1000 \times (0.03)^2}{(4\pi \times 40 \times 10^3)^2} = 1.425 \times 10^{-6} \text{ W}$$

$$= 1.425 \text{ μW}$$

(b) $L_{sys} = 10$ dB:

$$P_r = P_t \frac{G_t G_r \lambda_0^2}{(4\pi R)^2} \frac{1}{L_{sys}}$$

Therefore

$$P_r = 0.1425 \ \mu W \qquad \blacksquare$$

8.3   SPACE LOSS

Space loss accounts for the loss due to the spreading of RF energy as it propagates through free space. As can be seen, the power density $(P_t/4\pi R^2)$ from an isotropic antenna is reduced by $1/R^2$ as the distance is increased. Consider an isotropic transmitting antenna and an isotropic receiving antenna, as shown in Fig. 8.3. Equation (8.5) becomes

$$P_r = P_t \left(\frac{\lambda_0}{4\pi R}\right)^2 \qquad (8.12)$$

since $G_r = G_t = 1$ for an isotropic antenna. The term space loss (SL) is defined by

$$\text{SL in ratio} = \frac{P_t}{P_r} = \left(\frac{4\pi R}{\lambda_0}\right)^2 \qquad (8.13)$$

$$\text{SL in dB} = 10 \log \frac{P_t}{P_r} = 20 \log\left(\frac{4\pi R}{\lambda_0}\right) \qquad (8.14)$$

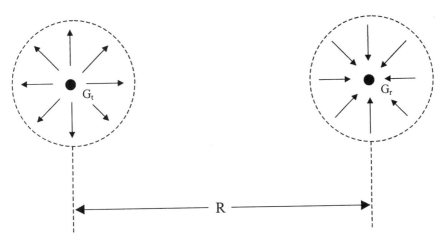

FIGURE 8.3   Two isotropic antennas separated by a distance R.

Example 8.2 Calculate the space loss at 4 GHz for a distance of 35,860 km.

Solution From Eq. (8.13),

$$\lambda_0 = \frac{c}{f} = \frac{3 \times 10^8}{4 \times 10^9} = 0.075 \text{ m}$$

$$\text{SL} = \left(\frac{4\pi R}{\lambda_0}\right)^2 = \left(\frac{4\pi \times 3.586 \times 10^7}{0.075}\right)^2$$

$$= 3.61 \times 10^{19} \quad \text{or } 196 \text{ dB} \qquad \blacksquare$$

8.4 LINK EQUATION AND LINK BUDGET

For a communication link, the Friis power transmission equation can be used to calculate the received power. Equation (8.5) is rewritten here as

$$P_r = P_t G_t G_r \left(\frac{\lambda_0}{4\pi R}\right)^2 \frac{1}{L_{\text{sys}}} \tag{8.15}$$

This is also called the link equation. System loss L_{sys} includes various losses due to, for example, antenna feed mismatch, pointing error, atmospheric loss, and polarization loss.

Converting Eq. (8.15) in decibels, we have

$$10 \log P_r = 10 \log P_t + 10 \log G_t + 10 \log G_r - 20 \log\left(\frac{4\pi R}{\lambda_0}\right) - 10 \log L_{\text{sys}} \tag{8.16a}$$

or

$$P_r = P_t + G_t + G_r - SL - L_{\text{sys}} \quad \text{(in dB)} \tag{8.16b}$$

From Eq. (8.16), one can set up a table, called a link budget, to calculate the received power by starting from the transmitting power, adding the gain of the transmitting antenna and receiving antenna, and subtracting the space loss and various losses.

Consider an example for a ground-to-satellite communication link (uplink) operating at 14.2 GHz as shown in Fig. 8.4 [1]. The ground station transmits an output power of 1250 W. The distance of transmission is 23,074 statute miles, or 37,134 km (1 statute mile = 1.609347219 km). The receiver in the satellite has a

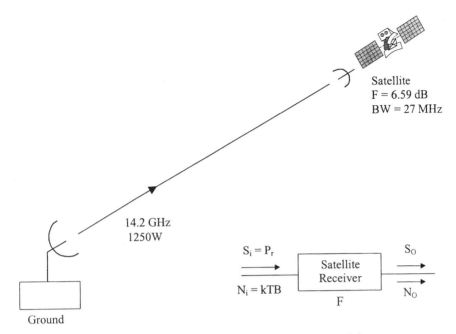

FIGURE 8.4   Ground-to-satellite communication uplink.

noise figure of 6.59 dB, and the bandwidth per channel is 27 MHz. At the operating frequency of 14.2 GHz, the free-space wavelength equals 0.0211 m. The space loss can be calculated by Eq. (8.14):

$$SL \text{ in dB} = 20 \log\left(\frac{4\pi R}{\lambda_0}\right) = 207.22 \text{ dB}$$

The following link budget chart can be set up:

Ground transmit power (P_t)	+30.97 dBW (1250 W)
Ground antenna feed loss	−2 dB
Ground antenna gain (G_t)	+ 54.53 dB
Ground antenna pointing error	−0.26 dB
Margin	−3 dB
Space loss	−207.22 dB
Atmospheric loss	−2.23 dB
Polarization loss	−0.25 dB
Satellite antenna feed loss	0 dB
Satellite antenna gain (G_r)	+37.68 dB
Satellite antenna pointing error	−0.31 dB
Satellite received power (P_r)	−92.09 dBW
	or − 62.09 dBm

The same P_r can be obtained by using Eq. (8.15) using L_{sys}, which includes the losses due to antenna feed, antenna pointing error, atmospheric loss, polarization loss, and margin. From the above table, L_{sys} is given by

$$L_{sys} = -2 \text{ dB} - 0.26 \text{ dB} - 3 \text{ dB} - 2.23 \text{ dB} - 0.25 \text{ dB} - 0.31 \text{ dB}$$
$$= -8.05 \text{ dB}$$

With the received power P_r at the input of the satellite receiver, one can calculate the receiver output SNR. From the definition of the noise factor, we have

$$F = \frac{S_i/N_i}{S_o/N_o} \tag{8.17}$$

The output SNR is given as

$$\frac{S_o}{N_o} = \frac{S_i}{N_i}\frac{1}{F} = \frac{S_i}{kTBF} = \frac{P_r}{kTBF} \tag{8.18}$$

For a satellite receiver with a noise figure of 6.59 dB and a bandwidth per channel of 27 MHz, the output SNR ratio at room temperature (290 K) used to calculate the standard noise power is

$$\frac{S_o}{N_o} \text{ in dB} = 10 \log \frac{S_o}{N_o} = 10 \log \frac{P_r}{kTBF}$$
$$= 10 \log P_r - 10 \log kTBF$$
$$= -92.09 \text{ dBW} - (-123.10 \text{ dBW})$$
$$= 31.01 \text{ dB or } 1262 \tag{8.19}$$

This is a good output SNR. The high SNR will ensure system operation in bad weather and with a wide temperature variation. The atmospheric loss increases drastically during a thunderstorm. The satellite receiver will experience fairly big temperature variations in space.

Example 8.3 At 10 GHz, a ground station transmits 128 W to a satellite at a distance of 2000 km. The ground antenna gain is 36 dB with a pointing error loss of 0.5 dB. The satellite antenna gain is 38 dB with a pointing error loss of 0.5 dB. The atmospheric loss in space is assumed to be 2 dB and the polarization loss is 1 dB. Calculate the received input power level and output SNR. The satellite receiver has a noise figure of 6 dB at room temperature. A bandwidth of 5 MHz is required for a channel, and a margin (loss) of 5 dB is used in the calculation.

Solution   First, the space loss is calculated:

$$\lambda_0 = c/f = 0.03 \text{ m} \qquad R = 2000 \text{ km}$$

$$\text{Space loss in dB} = 20 \log\left(\frac{4\pi R}{\lambda_0}\right) = 178.5 \text{ dB}$$

The link budget table is given below:

Ground transmit power	+21.1 dBW (or 128 W)
Ground antenna gain	+36 dB
Ground antenna pointing error	−0.5 dB
Space loss	−178.5 dB
Atmospheric loss	−2 dB
Polarization loss	−1 dB
Satellite antenna gain	+38 dB
Satellite antenna pointing error	−0.5 dB
Margin	−5 dB
Received signal power	−92.4 dBW
	or − 62.4 dBW

The output S_o/N_o in decibels is given by Eq. (8.19):

$$\frac{S_o}{N_o} \text{ in dB} = 10 \log \frac{P_r}{kTBF}$$

$$= 10 \log P_r - 10 \log kTBF$$

$$= -92.4 \text{ dBW} - (-130.99 \text{ dBW})$$

$$= 38.59 \text{ dB}$$

The same results can be obtained by using Eqs. (8.15) and (8.11) rewritten below:

$$P_r = P_t G_t G_r \left(\frac{\lambda_0}{4\pi R}\right)^2 \frac{1}{L_{\text{sys}}}$$

$$\frac{S_o}{N_o} = \frac{P_t G_t G_r}{kTBF \, L_{\text{sys}}} \left(\frac{\lambda_0}{4\pi R}\right)^2$$

Now

$$P_t = 128 \text{ W} \qquad\qquad G_t = 36 \text{ dB} = 3981$$
$$G_r = 38 \text{ dB} = 6310 \qquad \lambda_0 = 0.03 \text{ m}$$
$$k = 1.38 \times 10^{-23} \text{ J/K} \qquad T = 290 \text{ K}$$
$$B = 5 \text{ MHz} = 5 \times 10^6 \text{ Hz} \qquad F = 6 \text{ dB} = 3.98$$

$$L_{sys} = 0.5 \text{ dB} + 2 \text{ dB} + 1 \text{ dB} + 0.5 \text{ dB} + 5 \text{ dB} = 9 \text{ dB} = 7.94$$
$$R = 2000 \text{ km} = 2 \times 10^6 \text{ m}$$

$$P_r = 128 \text{ W} \times 3981 \times 6310 \times \left(\frac{0.03 \text{ m}}{4\pi \times 2 \times 10^6 \text{ m}}\right)^2 \frac{1}{7.94}$$

$$= 5.770 \times 10^{-10} \text{ W}$$

$$= -92.39 \text{ dBW}$$

$$\frac{S_o}{N_o} = \frac{128 \text{ W} \times 3981 \times 6310}{1.38 \times 10^{-23} \text{ W/sec/K} \times 290 \text{ K} \times 5 \times 10^6/\text{sec} \times 3.98 \times 7.94}$$
$$\times \left(\frac{0.03 \text{ m}}{4\pi \times 2 \times 10^6 \text{ m}}\right)^2$$

$$= 7245 \text{ or } 38.60 \text{ dB}$$

8.5 EFFECTIVE ISOTROPIC RADIATED POWER AND G/T PARAMETERS

The effective isotropic radiated power (EIRP) is the transmitted power that would be required if the signal were being radiated equally into all directions instead of being focused. Consider an isotropic antenna transmitting a power P_t' and a directional antenna transmitting P_t as shown in Fig. 8.5, with a receiver located at a distance R from the antennas. The received power from the isotropic antenna is

$$P_r' = \frac{P_t'}{4\pi R^2} A_{er} = \frac{P_t'}{4\pi R^2} \frac{G_r \lambda_0^2}{4\pi} = P_t' G_r \left(\frac{\lambda_0}{4\pi R}\right)^2 \qquad (8.20)$$

The received power from a directive antenna is, from Eq. (8.5),

$$P_r = P_t G_t G_r \left(\frac{\lambda_0}{4\pi R}\right)^2 \qquad (8.21)$$

where

$$P_r' = P_r, \qquad P_t' = P_t G_t = \text{EIRP} \qquad (8.22)$$

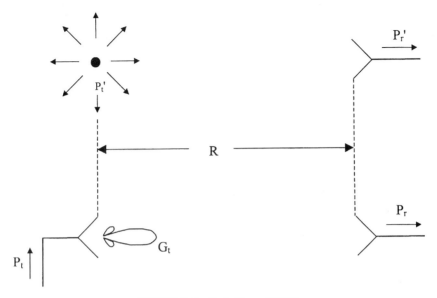

FIGURE 8.5 Definition of EIRP.

Thus EIRP is the amount of power that would be transmitted by an isotropic radiator given the measured receiver power. In a communication system, the larger the EIRP, the better the system. Therefore, we have

$$\text{EIRP} \equiv P_t G_t = \frac{P_r}{G_r} \left(\frac{4\pi R}{\lambda_0} \right)^2 \tag{8.23}$$

Example 8.4 A transmitting antenna has a gain of 40 dB and transmits an output power level of 100 W. What is the EIRP?

Solution

$$P_t = 100 \text{ W} = 20 \text{ dBW}$$
$$G_t = 40 \text{ dB} = 10,000$$
$$\text{EIRP} = P_t G_t = 1 \times 10^6 \text{ W or } 60 \text{ dBW} \qquad \blacksquare$$

The *G/T* parameter is a figure of merit commonly used for the earth station to indicate its ability to receive weak signals in noise, where *G* is the receiver antenna gain (G_r) and *T* is the system noise temperature (T_s).

The output SNR for a communication is given in Eq. (8.11) and rewritten here as

$$\frac{S_o}{N_o} = \frac{P_t G_t G_r}{kTBF \, L_{\text{sys}}} \left(\frac{\lambda_0}{4\pi R} \right)^2 \tag{8.24}$$

Substituting EIRP, space loss, and the G/T parameter into the above equation, we have

$$\frac{S_o}{N_o} = \frac{(\text{EIRP})(G_r/T_s)T_s}{(\text{space loss})kTBF\,L_{\text{sys}}} \tag{8.25}$$

It can be seen from the above equation that the output SNR ratio is proportional to EIRP and G_r/T_s but inversely proportional to the space loss, bandwidth, receiver noise factor, and system loss.

8.6   RADIO/MICROWAVE LINKS

A radio/microwave link is a point-to-point communication link using the propagation of electromagnetic waves through free space. Very and ultrahigh frequencies (VHF, UHF) are used extensively for short-range communications between fixed points on the ground. Frequently LOS propagation is not possible due to the blockage of buildings, trees, or other objects. Scattering and diffraction around the obstacles will be used for receiving with higher loss, as shown in Fig. 8.6a. Other examples of single-stage radio/microwave links are cordless phones, cellular phones, pager systems, CB radios, two-way radios, communications between aircraft and ships, and communications between air and ground. Figure 8.6b shows some examples.

For long-range communication between fixed points, microwave relay systems are used. If the points are close enough such that the earth's curvature can be neglected and if there is no obstruction, the link can be established as a single stage. For longer distances, multiple hops or a relay system is required, as shown in Fig. 8.7a. The relay station simply receives, amplifies, and retransmits the signals. In mountainous areas, it is possible to use a passive relay station located between the two relay stations. The passive station consists of a reflecting mirror in the direct view of each station, as shown if Fig. 8.7b. Many telephone and TV channels are transmitted to different areas using microwave links.

Tropospheric scattering is used for microwave links for long-distance communications (several hundred miles apart). There is no direct LOS between the stations. Microwaves are scattered by the nonhomogeneous regions of the troposphere at altitudes of 20 km, as shown in Fig. 8.8. Since only a small fraction of energy is scattered to the receiving antenna, high transmitting power, low receiver noise, and high antenna gain are required for reasonable performance. The operation can be improved by frequency diversity using two frequencies separated by 1% and by space diversity using two receiving antennas separated by a hundred wavelengths. Several received signals will be obtained with uncorrelated variation. The strongest can then be selected. Even longer distances can be established using ionospheric reflection. Layers of the ionosphere are located at altitudes of approximately 100 km.

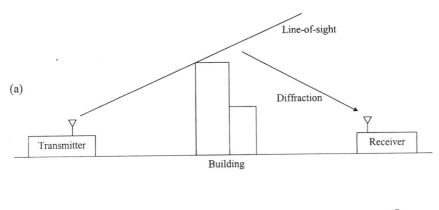

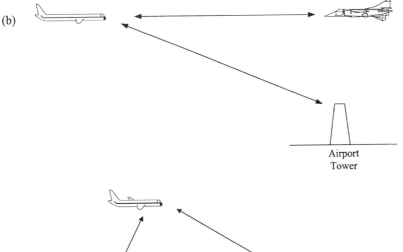

FIGURE 8.6   Radio/microwave links.

Communication links over thousands of miles can be established by using successive reflections of the waves between the earth's surface and the ionospheric layers. This method is still used by amateur radio and maritime communications.

8.7   SATELLITE COMMUNICATION SYSTEMS

In 1959, J. R. Pierce and R. Kompfner described the transoceanic communication by satellites [2]. Today, there are many communication satellites in far geosynchronous orbit (GEO) (from 35,788 to 41,679 km), near low earth orbit (LEO) (from 500 to

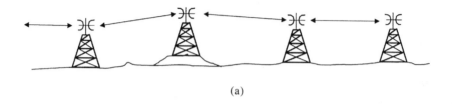

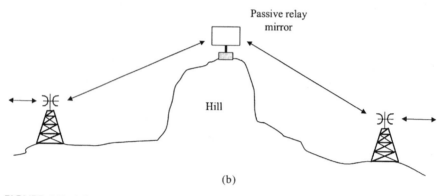

(a)

(b)

FIGURE 8.7   Microwave relay systems: (*a*) relay stations; (*b*) relay stations with a passive relay mirror.

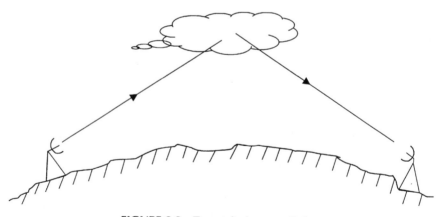

FIGURE 8.8   Tropospheric scatter link.

2000 km), and at medium-altitude orbit (MEO), in between the GEO and LEO orbits [3, 4]. Technological advances have resulted in more alternatives in satellite orbits, more output power in transmitters, lower noise in receivers, higher speed in modulation and digital circuits, and more efficient solar cells. Satellite communications have provided reliable, instant, and cost-effective communications on regional, domestic, national, or global levels. Most satellite communications use a fixed or mobile earth station. However, recent developments in personal communications have extended the direct satellite link using a hand-held telephone or laptop computer.

A simple satellite communication link is shown in Fig. 8.9. The earth station A transmits an uplink signal to the satellite at frequency f_U. The satellite receives, amplifies, and converts this signal to a frequency f_D. The signal at f_D is then transmitted to earth station B. The system on the satellite that provides signal receiving, amplification, frequency conversion, and transmitting is called a repeater or transponder. Normally, the uplink is operating at higher frequencies because higher frequency corresponds to lower power amplifier efficiency. The efficiency is less important on the ground than on the satellite. The reason for using two different uplink and downlink frequencies is to avoid the interference, and it allows simultaneous reception and transmission by the satellite repeaters. Some commonly used uplink and downlink frequencies are listed in Table 8.1. For example, at the C-band, the 4-GHz band (3.7–4.2 GHz) is used for downlink and the 6-GHz band (5.925–6.425 GHz) for uplink.

The repeater enables a flow of traffic to take place between several pairs of stations provided a multiple-access technique is used. Frequency division multiple access (FDMA) will distribute links established at the same time among different frequencies. Time division multiple access (TDMA) will distribute links using the same frequency band over different times. The repeater can distribute thousands of telephone lines, many TV channels, and data links. For example, the INTELSAT repeater has a capacity of 1000 telephone lines for a 36-MHz bandwidth.

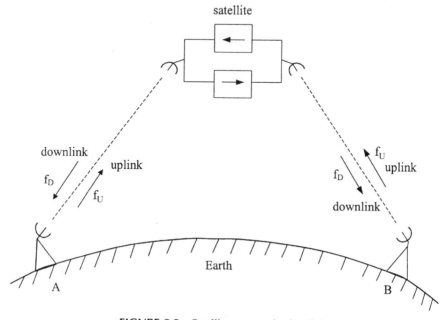

FIGURE 8.9 Satellite communication link.

TABLE 8.1   Commerical Satellite Communication Frequencies

Band	Uplink Frequency (GHz)	Downlink Frequency (GHz)
L	1.5	1.6
C	6	4
X	8.2	7.5
Ku	14	12
Ka	30	20
Q	44	21

The earth stations and satellite transponders consist of many RF and microwave components. As an example, Fig. 8.10 shows a simplified block diagram operating at the Ku-band with the uplink at 14–14.5 GHz and downlink at 11.7–12.2 GHz. The earth terminal has a block diagram shown in Fig. 8.11. It consists of two upconverters converting the baseband frequency of 70 MHz to the uplink frequency. A power amplifier (PA) is used to boost the output power before transmitting. The received signal is amplified by a low-noise RF amplifier (LNA) before it is downconverted to the baseband signal. The block diagram for the transponder on the satellite is shown in Fig. 8.12. The transponder receives the uplink signal (14–14.5 GHz). It amplifies the signal and converts the amplified signal to the downlink frequencies (11.7–12.2 GHz). The downlink signal is amplified by a power amplifier before transmitting. A redundant channel is ready to be used if any component in the regular channel is malfunctional. The redundant channel consists of the same components as the regular channel and can be turned on by a switch.

Figure 8.13 shows an example of a large earth station used in the INTELSAT system [5]. A high-gain dish antenna with Cassegrain feed is normally used. Because of the high antenna gain and narrow beam, it is necessary to track the satellite accurately within one-tenth of a half-power beamwidth. The monopulse technique is commonly used for tracking. The high-power amplifiers (HPAs) are either traveling-wave tubes (TWTs) or klystrons, and the LNAs are solid-state devices such as MESFETs. Frequency division multiple access and TDMA are generally used for modulation and multiple access of various channels and users.

8.8 MOBILE COMMUNICATION SYSTEMS AND WIRELESS CELLULAR PHONES

Mobile communication systems are radio/wireless services between mobile and land stations or between mobile stations. Mobile communication systems include maritime mobile service, public safety systems, land transportation systems, industrial systems, and broadcast and TV pickup systems. Maritime mobile service is between ships and coast stations and between ship stations. Public safety systems

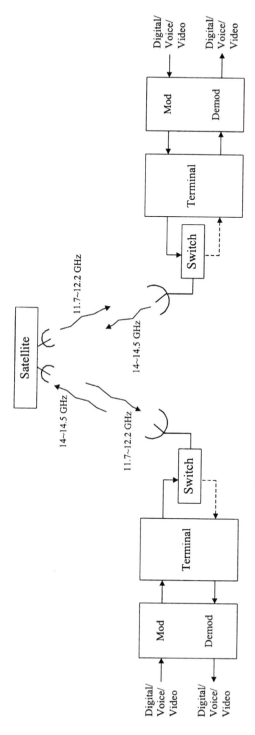

FIGURE 8.10 Simplified Ku-band satellite link.

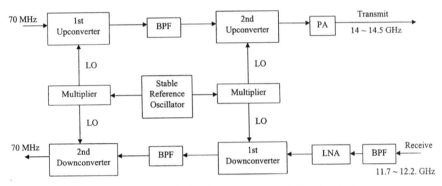

FIGURE 8.11   Block diagram for earth station terminal.

include police, fire, ambulance, highway, forestry, and emergency services. Land transportation systems cover the communications used by taxis, buses, trucks, and railroads. The industrial systems are used for communications by power, petroleum, gas, motion picture, press relay, forest products, ranchers, and various industries and factories.

Frequency allocations for these services are generally in HF, VHF, and UHF below 1 GHz. For example, the frequency allocations for public safety systems are [6] 1.605–1.750, 2.107–2.170, 2.194–2.495, 2.505–2.850, 3.155–3.400, 30.56–32.00, 33.01–33.11, 37.01–37.42, 37.88–38.00, 39.00–40.00, 42.00–42.95, 44.61–46.60, 47.00–47.69, 150.98–151.49, 153.73–154.46, 154.62–156.25, 158.7–159.48, 162.00–172.40, 453.00–454.00, and 458.0–459.0 MHz. The frequency allocations for the land transportation services are 30.56–32.00, 33.00–33.01, 43.68–44.61, 150.8–150.98, 152.24–152.48, 157.45–157.74, 159.48–161.57, 452.0–453.0, and 457.0–458.0 MHz.

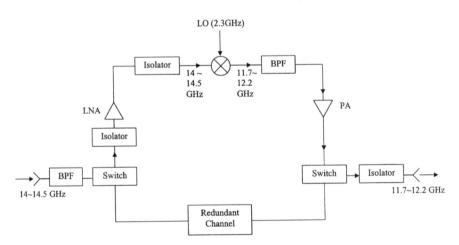

FIGURE 8.12   Block diagram for satellite transponder.

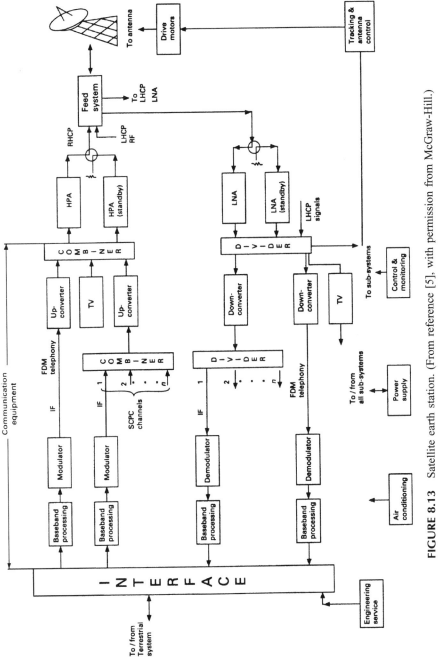

FIGURE 8.13 Satellite earth station. (From reference [5], with permission from McGraw-Hill.)

Land mobile communications have a long history. In the 1920s, one-way broadcasts were made to police cars in Detroit. The system was expanded to 194 municipal police radio stations and 5000 police cars in the 1930s. During World War II, several hundred thousand portable radio sets were made for military use. In the late 1940s, Bell System proposed the cellular concept. Instead of the previously used "broadcast model of a high-power transmitter," placed at a high elevation covering a large area, the new model used a low-power transmitter covering a small area called a "cell." Each cell has a base station that communicates with individual users. The base stations communicate to each other through a switching office and, from there, to satellites or the outside world. Figure 8.14 shows the concept of cells. There is a base station in each cell, and the actual cell shape may not be hexagonal. The system has the following features:

1. Since each cell covers a small area, low-power transmitters can be used in the base stations.
2. Frequencies can be reused with a sufficient separation distance between two cells. For example, the cells in Fig. 8.14 using the same letters (A, B, C, D, ...) are in the same frequency.
3. Large cells can be easily reduced to small cells over a period of time through splitting when the traffic is increased.
4. The base station can pass a call to other stations without interruption (i.e., hand-off and central control).

The first-generation cellular telephone system that started in the mid-1980s used analog modulation (FM). The second-generation system used digital modulation and TDMA. Some recent systems use code division multiple access (CDMA) to increase the capacity, especially in big cities. Table 8.2 summarizes the analog and digital cellular and cordless phone services. The information shown is just an example since the technology has changed very rapidly. Digital cellular phone systems offer greater user capacity, improved spectral efficiency, and enhanced voice quality and security. In Europe, the Global System for Mobile Communication (GSM) is a huge, rapidly expanding system. A typical GSM 900 (operating at 900 MHz) cell can be located up to a 35 km radius. GSM uses TDMA or FDMA operation.

8.9 PERSONAL COMMUNICATION SYSTEMS AND SATELLITE PERSONAL COMMUNICATION SYSTEMS

Personal communication systems, personal communication networks (PCNs), or local multipoint distribution service (LMDS) operate at higher frequencies with wider bandwidths. The systems offer not only baseline voice services like cellular phones but also voice-mail, messaging, database access, and on-line services, as shown in Fig. 8.15. Table 8.3 shows the frequency allocations for PCSs designated by the Federal Communications Commission.

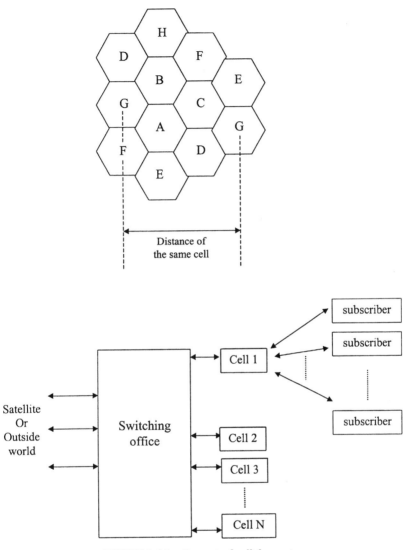

FIGURE 8.14 Concept of cellular systems.

The direct link between satellites and PCSs can provide data and voice communications anywhere in the world, even in the most remote regions of the globe. At least six satellite systems are under development for wireless personal voice and data communications using a combination of wireless telephones, wireless modems, terrestrial cellular telephones, and satellites. Many companies and consortia have invested billions of dollars to launch satellites capable of providing paging, voice, data, fax, and video conferencing worldwide. A few examples are given in Tables 8.4 and 8.5.

TABLE 8.2 Analog and Digital Cellular and Cordless Phone Services

Standard	Analog Cellular Telephones			Analog Cordless Telephones		
	Advanced Mobile Phone Service (AMPS)	Total Access Communication System (TACS)	Nordic Mobile Telephone (NMT)	Cordless Telephone (CTO)	Japanese Cordless Telephone (JCT)	Cordless Telephone 1 (CTI/CTI+)
Mobile frequency range (MHz)	Rx: 869–894, Tx: 824–849	ETACS: Rx: 916–949, Tx: 871–904 NTACS: Rx: 860–870, Tx: 915–925	NMT-450: Rx: 463–468, Tx: 453–458 NMT-900: Rx: 935–960, Tx: 890–915	2/48 (U.K.), 26/41 (France), 30/39 (Australia), 31/40 (The Netherlands, Spain), 46/49 (China, S. Korea, Taiwan, US), 48/74 (China)	254–380	CTI: 915/960, CTI+: 887/932
Multiple-access method	FDMA	FDMA	FDMA	FDMA	FDMA	FDMA
Duplex method	FDD	FDD	FDD	FDD	FDD	FDD
Number of channels	832	ETACS: 1000, NTACS: 400	NMT-450: 200, NMT-900: 1999	10, 12, 15 or 20	89	CTI: 40, CTI+: 80
Channel spacing	30 kHz	ETACS: 25 kHz, NTACS: 12.5 kHz	NMT-450: 25 kHz, NMT-900: 12.5 kHz	40 kHz	12.5 kHz	25 kHz
Modulation	FM	FM	FM	FM	FM	FM
Bit rate	n/a	n/a	n/a	n/a	n/a	n/a

TABLE 8.2 (*Continued*)

	Digital Cellular Telephones				Digital Cordless Telephones/PCN			
Standard	North American Digital Cellular (IS-54)	North American Digital Cellular (IS-95)	Global System for Mobile Communications (GSM)	Personal Digital Cellular (PDC)	Cordless Telephone 2 (CT2/CT2+)	Digital European Cordless Telephone (DECT)	Personal Handy Phone System (PHS)	DCS 1800
Mobile frequency range (MHz)	Rx: 869–894, Tx: 824–849	Rx: 869–894, Tx: 824–849	Rx: 935–960, Tx: 890–915	Rx: 810–826, Tx: 940–956; Rx: 1429–1453, Tx: 1477–1501	CT2: 864/868: CT2+: 930/931, 940/941	1880–1990	1895–1907	Rx: 1805–1880, Tx: 1710–1785
Multiple access method	TDMA/FDM	CDMA/FDM	TDMA/FDM	TDMA/FDM	TDMA/FDM	TDMA/FDM	TDMA/FDM	TDMA/FDM
Duplex method	FDD	FDD	FDD	FDD	TDD	TDD	TDD	FDD
Number of channels	832 (3 users/ channel)	20 (798 users/ channel)	124 (8 users/ channel)	1600 (3 users/ channel)	40	10 (12 users/ channel)	300 (4 users/ channel)	750 (16 users/ channel)
Channel spacing	30 kHz	1250 kHz	200 kHz	25 kHz	100 kHz	1.728 MHz	300 kHz	200 kHz
Modulation	π/4 DQPSK	BPSK/ OQPSK	GMSK (0.3 Gaussian filter)	π/4 DQPSK	GFSK (0.5 Gaussian filter)	GFSK (0.5 Gaussian filter)	π/4 DQPSK	GMSK (0.3 Gaussian filter)
Bit rate	48.6 kb/sec	1.288 Mb/sec	270.833 kb/sec	42 kb/sec	72 kb/sec	1.152 Mb/sec	384 kb/sec	270.833 kb/sec

Source: From reference [7], with permission from IEEE.

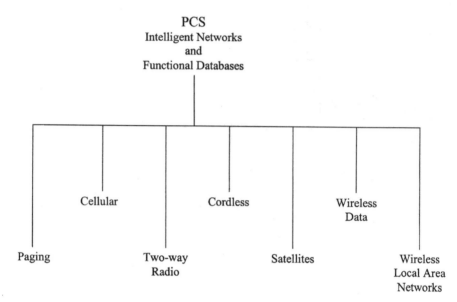

FIGURE 8.15   Personal communication systems.

With the combination of wireless telephones, wireless modems, terrestrial cellular telephones, WLANs, and satellites, the ultimate vision for PCSs is a wireless go-anywhere communicator [8]. In general, outdoor communications would be handled by PCS carriers connected to public voice and data networks through telephone, cable, and satellite media. The PCS microcells and WLANs with base stations could be installed indoors. Figure 8.16 shows an artist's picture of this vision [8]. The WLAN is for wireless indoor radio communication services or high-data-rate communications. The system employs a central microcell hub (base station) that services cordless phones and computer workstations whose transceivers are networking with the hub through wireless communications. Since the cell is small, inexpensive low-power base stations can be used. Several WLAN microcells are shown in Fig. 8.16. Table 8.6 shows some WLANs frequency allocations. One

TABLE 8.3   PCS Frequency Allocations

Channel Block	Frequency (MHz)	Service Area
A (30 MHz)	1850–1865/1930–1945	Major trading areas
B (30 MHz)	1865–1880/1945–1960	Major trading areas
C (20 MHz)	1880–1890/1960–1970	Basic trading areas
D (10 MHz)	2130–2135/2180–2185	Basic trading areas
E (10 MHz)	2135–2140/2185–2190	Basic trading areas
F (10 MHz)	2140–2145/2190–2195	Basic trading areas
G (10 MHz)	2145–2150/2195–2200	Basic trading areas

Source: Federal communications commission.

TABLE 8.4 Satellite Personal Communication Systems Under Development

	Globalstar	Teledesic	Iridium	American Mobile Satellite Corp.	Spaceway	Odyssey
Headquarters	Palo Alto, CA	Kirkland, WA	Washington, DC	Reston, VA	El Segundo, CA	Redondo Beach, CA
Investors	Loral Corp., Qualcomm, Alcatel (France), Dacom Corp. (S. Korea), and Deutsche Aerospace (Germany) and others	Startup company backed with funding from William Gates, chairman of Microsoft, and Craig McCaw, chairman of McCaw Cellular	Motorola, Sprint, STET (Italy), Bell Canada, DDI (Japan)	Hughes Communications, McCaw Cellular, Mtel, Singapore Telecomm	Wholly owned and operated by Hughes Communications	TRW, Teleglobe of Canada
Estimated cost to build	$1.8 billion	$9 billion	$3.4 billion	$550 million	$660 million	$1.3 billion
Description	Worldwide voice, data, fax, and paging services using 48 LEO satellites	A worldwide network of 840 satellites will offer voice, data, fax and two-way video communications	66 satellites will offer worldwide voice, data, fax, and paging services	Satellite network will provide voice, data, fax, and two-way messaging throughout North America, targeting customers in regions not served by cellular systems	Dual-satellite system offering voice, data, and two-way videoconferencing in North America	12 satellite system offering voice, data, and fax services

Source: From reference [7], with permission from IEEE.

TABLE 8.5 System Parameters of Several Satellite Personal Communication Systems

System	Iridium (Motorola)	Odyssey (TRW)	Globalstar (Loral and Qualcomm)
No. of satellites	66	12	48
Class	LEO	MEO	LEO
Lifetime in years	5	15	7.5
Orbit altitude (km)	781	10,354	1390
Orientation	Circular	Circular	Circular
Initial geographical coverage	Global	CONUS, Offshore U.S., Europe, Asia/Pacific	CONUS
Service markets	Cellular like voice, mobile FAX, paging, messaging, data transfer	Cellular like voice, mobile FAX, paging, messaging, data transfer	Cellular like voice, mobile FAX, paging, messaging, data transfer
Voice cost/minute	$3.00	$0.65	$0.30
User terminal types	Hand-held, vehicular, transportable	Hand-held, vehicular, transportable	Hand-held, vehicular, transportable
Estimated cost	$3000	$250–500	$500–700
Wattage	0.4	0.5	1
Uplink bands	L-band (1616.5–1625.5 MHz)	L-band (1610.0–1626.5 MHz)	L-band (1610.0–1626.5 MHz)
Downlink bands	L-band	S-band (2483.5–2500 MHz)	S-band (2483.5–2500 MHz)
Methods of access	FDMA	CDMA	CDMA

Source: From reference [7], with permission from IEEE.

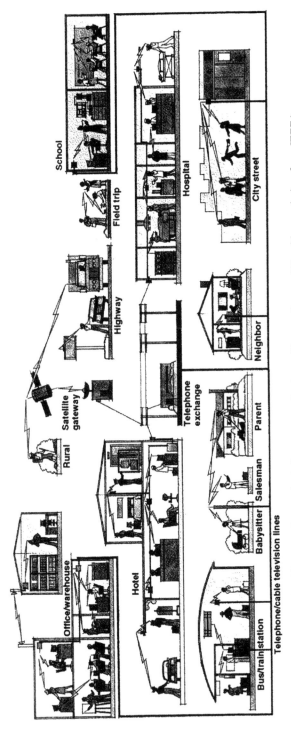

FIGURE 8.16 Personal wireless communications. (From reference [8], with permission from IEEE.)

269

TABLE 8.6   Wireless Local-Area Network Frequency Bands

Unlicensed	Licensed	No license
ISM	Narrowband	Infrared
0.902–0.928 GHz	Microwave	(IR)
2.400–2.483 GHz	18.82–18.87 GHz	No FCC regulation
5.725–5.825 GHz	19.16–19.21 GHz	
Spread-Spectrum $P_0 < 1$ watt		

Source: From reference [7], with permission from IEEE.

application of WLANs is wireless offices. With the rapid pace of office relocation and changing assignments, WLANs eliminate the need for running wires over, under, and through offices.

PROBLEMS

8.1   (a) What is the required receiver antenna gain in decibels for a CW communication system with a transmitted power of 1 kW operating at 30 GHz? The transmit antenna is a dish with a diameter of 1 m and 55% efficiency. The receiver requires an input signal S_{min} of 10^{-6} W for good reception. The maximum range of the system is designed to be 100 km for the ideal case. Neglect the system losses. (b) What is the range if the transmitted power is doubled to 2 kW? What is the range if the transmitted power is 200 kW?

8.2   A communication system has a transmitter with an output power of 1000 W at 10 GHz. The transmit antenna is a dish with a diameter of 4 m. The receive antenna is a dish with a diameter of 2 m. Assume that the antennas have 55% efficiencies. The distance of communication is 50 km. The receiver has a bandwidth of 25 MHz and noise figure of 10 dB operating at room temperature (290 K), (a) Calculate the received power in watts. (b) Calculate the SNR ratio at the output port of the receiver. Assume that the overall system loss is 6 dB.

8.3   A mobile communication system has the following specifications:

Transmitter power output $= 10$ W
Transmitter antenna gain $= 10$ dB
Overall system loss $= 10$ dB
Receiver noise figure $= 4$ dB
Operating frequency $= 2$ GHz

Receiver antenna gain $= 2\,dB$

Temperature $= 290\,K$

Receiver bandwidth $= 3\,MHz$

Required minimum output SNR for the receiver $= 15\,dB$

Calculate the maximum range in km.

8.4 In a communication system, if the receiver output SNR is increased, determine if the EIRP is (increased, decreased), the range is (increased, decreased), the receiver noise is (increased, decreased), and the operating bandwidth is (increased, decreased). What is the EIRP for the transmitter given in Problem 8.3?

8.5 In Section 8.4, the link budget for a satellite communication uplink case was discussed. For a downlink case, the frequency is 12.1 GHz. The satellite antenna gain at this frequency is 36.28 dB. The satellite tube's output power is 200 W. The atmospheric loss at this frequency is 1.52 dB. The ground terminal antenna gain is 53.12 dB, and the ground terminal receiver noise figure is 4.43 dB. All other parameters are the same as the uplink case. Calculate (a) the space loss in decibels, (b) the received power in watts, (c) the received carrier-to-noise power ratio (at the receiver output) in decibels, and (d) the EIRP in decibels relative to 1 W (dBW). (e) If the bandwidth is 20 MHz, design the terminal antenna gain to maintain the same SNR ratio as in part (c). In (b), give a table showing the link budget.

8.6 A ground-to-satellite communication link is operating at 15 GHz. The satellite is located in the LEO at a distance of 2000 km from the ground station. Quadriphase shift keying (QPSK) modulation is used to transmit a data rate of 10 Mbits/sec, which is equivalent to a 5-MHz bandwidth. The satellite receiver has a noise figure of 8 dB, an antenna gain of 41 dB, and an antenna pointing error loss of 0.5 dB. The ground transmitter has an output power of 82 W, an antenna gain of 38 dB, and an antenna pointing error loss of 0.5 dB. The atmospheric loss in fair weather is 3 dB; polarization loss is 1 dB. A margin of 5 dB is used. (a) Set up a link budget table and calculate the received power. (b) Calculate the receiver output SNR ratio. (c) What is the receiver output SNR ratio if the atmospheric loss is 20 dB in a thunderstorm?

8.7 A communication link has the link budget shown below. The distance of communication is 50 km. The operating frequency is 3 GHz. The receiver has a noise figure of 5 dB, a bandwidth of 10 MHz, and an operating temperature of 290 K. Calculate (a) the space loss in decibels, (b) the received power in watts, (c) the received carrier-to-noise power ratio in decibels (at the receiver output), and (d) the EIRP in watts. (e) What is the transmit power in watts if the transmit antenna is a dipole antenna with a gain of 2 dB and we want to maintain the same EIRP and received power?

Transmitting power (100 W)	___dBW
Feed loss	−1 dB
Transmitting antenna gain	40 dB
Antenna pointing error	−1 dB
Space loss	___dB
Atmospheric loss	−2 dB
Polarization loss	−0.5 dB
Receiving feed loss	−1 dB
Receiving antenna gain	20 dB
Receiving antenna pointing error	+) −1 dB

8.8   A microwave link is set up to communicate from a city to a nearby mountain top, as shown in Fig. P8.8. The distance is 100 km, and the operating frequency is 10 GHz. The receiver has a noise figure of 6 dB, a bandwidth of 30 MHz, and an operating temperatuare of 290 K. Calculate (a) the received power in watts and (b) the received carrier-to-noise power ratio in decibels at the receiver output port. (c) Repeat (b) if there is a thunderstorm that gives an additional loss of 5 dB/km for a region of 5 km long between the transmitter and receiver.

Transmitting power (1000 W)	___dBW
Feed loss	−1.5 dB
Transmitting antenna gain	45 dB
Antenna pointing error	−1 dB
Space loss	___dB
Atmospheric loss	−2 dB
Polarization loss	−0.5 dB
Receiving feed loss	−1.5 dB
Receiving antenna gain	45 dB
Receiving antenna pointing error	+) −2 dB

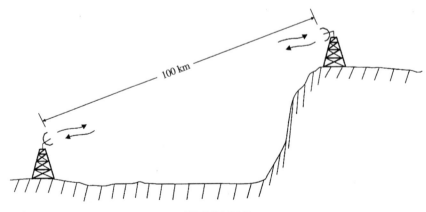

FIGURE P8.8

8.9   A microwave link is used to connect two communication towers, as shown in Fig. P8.9. The distance is 100 km, and the operating frequency is 10 GHz. The receiver has a noise figure of 5 dB and a bandwidth of 10 MHz and operates at room temperature (290 K). Calculate (a) the received power in watts, (b) the received carrier-to-noise power ratio in decibels at the input of the receiver, (c) the received carrier-to-noise ratio in decibels at the output of the receiver, (d) the transmitter EIRP in watts, and (e) the receiver output SNR ratio if a low-gain dipole antenna (gain = 2 dB) is used in the receiver.

Transmitting power (100 W)	___dBW
Feed loss	−1 dB
Transmitting antenna gain	35 dB
Antenna pointing error	−1 dB
Space loss	___dB
Atmospheric loss	−2 dB
Polarization loss	−1 dB
Receiving feed loss	−2 dB
Receiving antenna gain	35 dB
Receiving antenna pointing error	+) −1 dB

FIGURE P8.9

REFERENCES

1. E. A. Wolff and R. Kaul, *Microwave Engineering and Systems Applications*, John Wiley & Sons, New York, 1988.

2. J. R. Pierce and R. Kumpfner, "Transoceanic Communication by Means of Satellites," *Proc. IRE*, Vol. 47, pp. 372–380, March 1959.

3. G. D. Gordon and W. L. Morgan, *Principles of Communications Satellites*, John Wiley & Sons, New York, 1993.

4. W. L. Morgan and G. D. Gordon, *Communications Satellite Handbook*, John Wiley & Sons, New York, 1989.

5. M. Richharia, *Satellite Communication Systems*, McGraw-Hill, New York, 1995.

6. F. Losee, *RF Systems, Components, and Circuits Handbook*, Artech House, Boston, 1997.

7. F. Ali and J. B. Horton, "Introduction to Special Issue on Emerging Commercial and Consumer Circuits, Systems, and Their Applications," *IEEE Trans. Microwave Theory Tech.*, Vol. 43, No. 7, pp. 1633–1637, 1995.

8. B. Z. Kobb, "Personal Wireless," *IEEE Spectrum*, pp. 20–25, June 1993.

CHAPTER NINE

Modulation and Demodulation

9.1 INTRODUCTION

Modulation is a technique of imposing information (analog or digital) contained in a lower frequency signal onto a higher frequency signal. The lower frequency is called the modulating signal, the higher frequency signal is called the carrier, and the output signal is called the modulated signal. The benefits of the modulation process are many, such as enabling communication systems to transmit many baseband channels simultaneously at different carrier frequencies without their interfering with each other. One example is that many users can use the same long-distance telephone line simultaneously without creating a jumbled mess or interference. The modulation technique also allows the system to operate at a higher frequency where the antenna is smaller.

Some form of modulation is always needed in an RF system to translate a baseband signal (e.g., audio, video, data) from its original frequency bandwidth to a specified RF frequency spectrum. Some simple modulation can be achieved by direct modulation through the control of the bias to the active device. A more common method is the use of an external modulator at the output of the oscillator or amplifier. Figure 9.1 explains the concept of modulation.

There are many modulation techniques, for example, AM, FM, amplitude shift keying (ASK), frequency shift keying (FSK), phase shift keying (PSK), biphase shift keying (BPSK), quadriphase shift keying (QPSK), 8-phase shift keying (8-PSK), 16-phase shift keying (16-PSK), minimum shift keying (MSK), and quadrature amplitude modulation (QAM). AM and FM are classified as analog modulation techniques, and the others are digital modulation techniques.

After modulation, the signal is amplified and radiated to free space by an antenna. The signal is then picked up by a receiver antenna at some distance away and is then

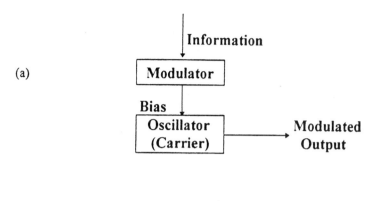

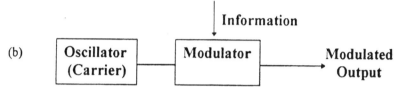

FIGURE 9.1 Different modulation schemes: (*a*) direct modulation; (*b*) external modulation.

amplified, downconverted, and demodulated to recover the original baseband signal (information).

9.2 AMPLITUDE MODULATION AND DEMODULATION

Amplitude and frequency modulation techniques are classified as analog modulation. They are old techniques, having been used for many years since the invention of the radio. Analog modulation uses the baseband signal (modulating signal) to vary one of three variables: amplitude A_c, frequency ω_c, or phase θ. The carrier signal is given by

$$v_c(t) = A_c \sin(\omega_c t + \theta) \tag{9.1}$$

The amplitude variation is AM, the frequency variation is FM, and the phase variation is PM. Phase modulation and FM are very similar processes and can be referred to as angle modulation.

The unique feature of AM is that the message of the modulated carrier has the same shape as the message waveform. Figure 9.2 illustrates the carrier, modulating, and modulated signals.

For simplicity, let a single audio tone be a modulating signal

$$v(t) = A_m \sin \omega_m t \tag{9.2}$$

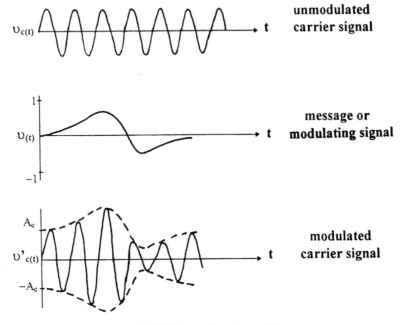

FIGURE 9.2   Signals in AM.

Although a sine wave is assumed, a more complex wave can be considered to be the sum of a set of pure sine waves.

The modulated signal can be written as

$$v_c'(t) = (A_c + A_m \sin \omega_m t) \sin \omega_c t$$

$$= A_c \left(1 + \frac{A_m}{A_c} \sin \omega_m t\right) \sin \omega_c t$$

$$= A_c(1 + m \sin \omega_m t) \sin \omega_c t \tag{9.3}$$

where

$$m = \frac{A_m}{A_c} = \frac{\text{peak value of modulating signal}}{\text{peak value of unmodulated carrier signal}}$$

where m is the modulation index, which sometimes is expressed in percentage as the percent of modulation. To preserve information without distortion would require m to be ≤ 1 or less than 100%. Figure 9.3 shows three cases of modulation: under-modulation ($m < 100\%$), 100% modulation, and overmodulation ($m > 100\%$).

Using a trigonometric identity, Eq. (9.3) can be rewritten as

$$v_c'(t) = A_c \sin \omega_c t + \tfrac{1}{2}(mA_c) \cos(\omega_c - \omega_m)t - \tfrac{1}{2}(mA_c) \cos(\omega_c + \omega_m)t \tag{9.4}$$

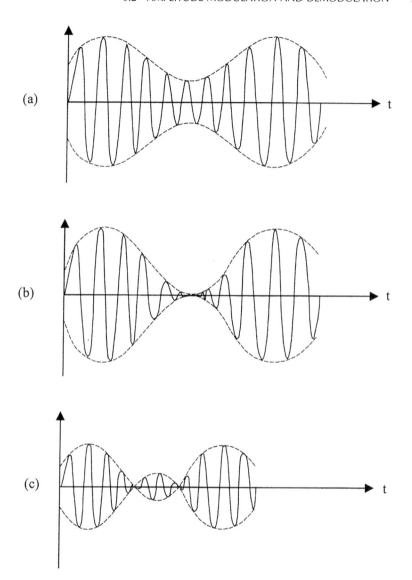

FIGURE 9.3 Degrees of modulatioin: (*a*) undermodulation; (*b*) 100% modulation; (*c*) overmodulation.

The modulated signal contains the carrier signal (ω_c), the upper sideband signal ($\omega_c + \omega_m$), and the lower sideband signal ($\omega_c - \omega_m$). This is quite similar to the output of a mixer.

A nonlinear device can be used to accomplish the amplitude modulation. Figure 9.4 shows examples using a modulated amplifier and a balanced diode modulator.

The demodulation can be achieved by using an envelope detector (described in Chapter 4) as a demodulator to recover the message [1].

Example 9.1   In an AM broadcast system, the total transmitted power is 2000 W. Assuming that the percent of modulation is 100%, calculate the transmitted power at the carrier frequency and at the upper and lower sidebands.

Solution   From Equation (9.4)

$$P_T = P_c + \tfrac{1}{4}m^2P_c + \tfrac{1}{4}m^2P_c = 2000 \text{ W}$$

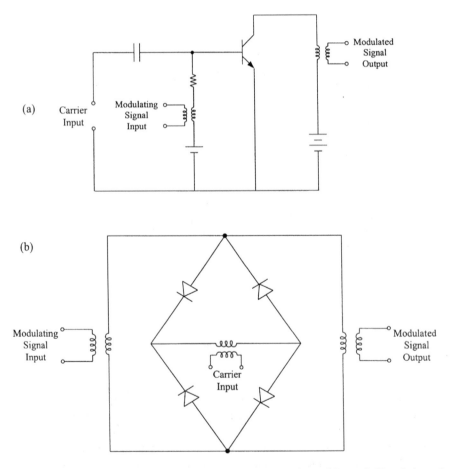

FIGURE 9.4   Amplitude modulation using (*a*) a modulated amplifier and (*b*) a balanced modulator.

Now $m = 1$, we have

$$1.5P_c = 2000 \text{ W} \qquad P_c = 1333.33 \text{ W}$$

Power in the upper sideband $= P_{\text{USB}} = \frac{1}{4}m^2 P_c = \frac{1}{4}P_c = 333.33 \text{ W}$

Power in the lower sideband $= P_{\text{LSB}} = \frac{1}{4}m^2 P_c = 333.33 \text{ W}$ ◼

9.3 FREQUENCY MODULATION

Frequency modulation is accomplished if a sinusoidal carrier, shown in Eq. (9.1), has its instantaneous phase $\omega_c t + \theta$ varied by a modulating signal. There are two possibilities: Either the frequency $\omega_c/2\pi$ or the phase θ can be made to vary in direct proportion to the modulating signal. The difference between FM and PM is not obvious, since a change in frequency must inherently involve a change in phase. In FM, information is placed on the carrier by varying its frequency while its amplitude is fixed.

The carrier signal is given by

$$v_c(t) = A_c \sin \omega_c t \tag{9.5}$$

The modulating signal is described as

$$v(t) = A_m \sin \omega_m t \tag{9.6}$$

The modulated signal can be written as

$$v'_c(t) = A_c \sin[2\pi(f_c + \Delta f \sin 2\pi f_m t)t] \tag{9.7}$$

The maximum frequency swing occurs when $\sin 2\pi f_m t = \pm 1$. Here Δf is the frequency deviation, which is the maximum change in frequency the modulated signal undergoes. The amplitude remains the same. A modulation index is defined as

$$m_f = \frac{\Delta f}{f_m} \tag{9.8}$$

The total variation in frequency from the lowest to the highest is referred to as carrier swing, which is equal to $2\,\Delta f$.

In the transmitter, frequency modulation can be achieved by using VCOs. The message or modulating signal will control the VCO output frequencies. In the receiver, the demodulator is used to recover the information. One example is to use a frequency discriminator (frequency detector) that produces an output voltage that is dependent on input frequency. Figure 9.5 shows a block diagram, a circuit schematic, and the voltage–frequency characteristics of a balanced frequency

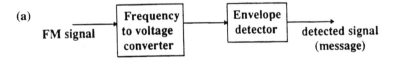

(a)

FM signal → Frequency to voltage converter → Envelope detector → detected signal (message)

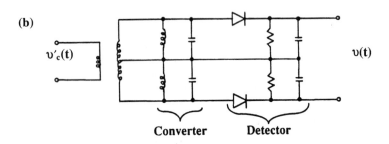

(b)

$v'_c(t)$ Converter   Detector $v(t)$

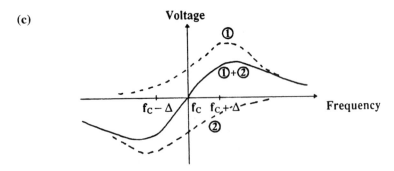

(c)

Voltage

①

①+②

$f_C - \Delta$ f_C $f_C + \Delta$ Frequency

②

FIGURE 9.5   Balanced frequency discrimination: (a) block diagram; (b) circuit schematic; (c) voltage–frequency characteristics.

discriminator. The circuit consists of a frequency-to-voltage converter and an envelope detector. The balanced frequency-to-voltage converter has two resonant circuits, one tuned above f_c and the other below. Taking the difference of these gives the frequency-to-voltage characteristics of an S-shaped curve. The conversion curve is approximately linear around f_c. Direct current is automatically canceled, bypassing the need for a DC block.

9.4   DIGITAL SHIFT-KEYING MODULATION

Most modern wireless systems use digital modulation techniques. Digital modulation offers many advantages over analog modulation: increased channel capability, greater accuracy in the presence of noise and distortion, and ease of handling. In digital communication systems, bits are transmitted at a rate of kilobits, megabits, or gigabits per second. A certain number of bits represent a symbol or a numerical number. The receiver then estimates which symbol was originally sent from the transmitter. It is largely unimportant if the amplitude or shape of the received signal is distorted as long as the receiver can clearly distinguish one symbol from the other. Each bit is either 1 or 0. The addition of noise and distortion to the signal makes it harder to determine whether it is 1 or 0. If the distortion is under a certain limit, the receiver will make a correct estimate. If the distortion is too large, the receiver may give a wrong estimate. When this happens, a BER is generated. Most wireless systems can tolerate a BER of 10^{-3} (1 in 1000) before the performance is considered unacceptable.

Amplitude shift keying, FSK, BPSK, QPSK, 8-PSK, 16-PSK, MSK, Gaussian MSK (GMSK), and QAM are classified as digital modulation techniques. A brief description of these modulation methods is given below.

In ASK modulation, the amplitude of the transmitted signal is turned "on" and "off," which corresponds to 1 or 0. This can easily be done by bias modulating an oscillator; that is, the oscillator is switched on and off by DC bias. Alternatively, a single-pole, single-throw $p-i-n$ or FET switch can be used as a modulator. Figure 9.6 shows the modulation arrangement for ASK. Demodulation can be obtained by a detector described in Chapter 4 [1, Ch. 6].

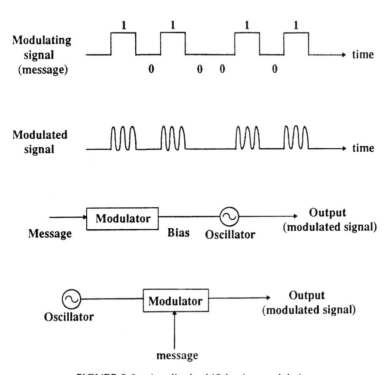

FIGURE 9.6 Amplitude shift keying modulation.

With FSK, when the modulating signal is 1, the transmitter transmits a carrier at frequency f_1; when the modulating signal is 0, the transmitting frequency is f_0. A VCO can be used to generate the transmitting signal modulated by the message. At the receiver, a frequency discriminator is used to distinguish these two frequencies and regenerate the original bit stream.

Minimum shift keying is the binary FSK with two frequencies selected to ensure that there is exactly an 180° phase shift difference between the two frequencies in a 1-bit interval. Therefore, MSK produces a maximum phase difference at the end of the bit interval using a minimum difference in frequencies and maintains good phase continuity at the bit transitions (see Fig. 9.7a [2]). Minimum shift keying is attractive

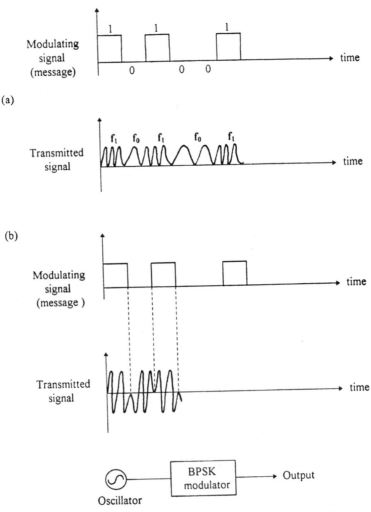

FIGURE 9.7   Modulation techniques: (a) MSK; (b) BPSK.

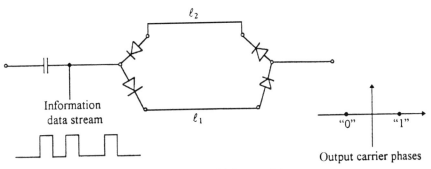

FIGURE 9.8   Biphase switch.

because it has a more compact spectrum and lower out-of-band emission as compared to FSK. Out-of-band emission can cause adjacent channel interference and can be further reduced by using filters. If a Gaussian-shaped filter is used, the modulation technique is called Gaussian MSK (GMSK).

In a PSK system, the carrier phase is switched between various discrete and equispaced values. For a BPSK system, the phase angles chosen are $0°$ and $180°$. Figure 9.7 shows the MSK and BPSK system waveforms for comparison. A switch can be used as a BPSK modulator. Figure 9.8 shows an example circuit. When the data are positive or "1," the signal passes path 1 with a length l_1. When the data are negative or "0," the signal goes through path 2 with a length l_2. If the electrical phase difference for these two paths is set equal to $180°$, we have a biphase switch/modulator. This is given by

$$\Delta\phi = \beta(l_1 - l_2) = \frac{2\pi}{\lambda_g}(l_1 - l_2) = 180° \tag{9.9}$$

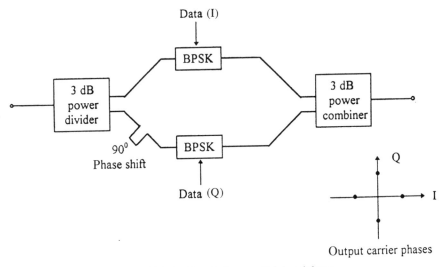

FIGURE 9.9   Quadriphase switch/modulator.

A QPSK modulator consists of two BPSK modulators, connected as shown in Fig. 9.9. A 90° phase shift made of a transmission line is used to introduce the 90° rotation between the outputs of the two BPSK switches. An output phase error of less than 3° and maximum amplitude error of 0.5 dB have been reported at 60 GHz using this circuit arrangement [3]. Quadrature PSK can transmit higher data rates, since two data streams can be transmitted simultaneously. Therefore, the theoretical bandwidth efficiency for QPSK is 2 bits per second per hertz (bps/Hz) instead of 1 bps/Hz for BPSK. Quadrature PSK transmits four (2^2) phases of 0°, 90°, 180°, and 270°. Two data streams can be transmitted simultaneously. The in-phase (I) data stream transmits 0° or 180° depending on whether the data are 1 or 0. The quadrature-phase (Q) data stream transmits 90° and 270°.

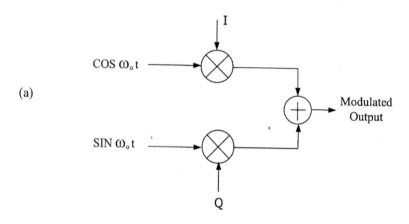

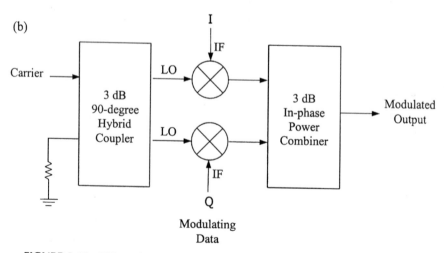

FIGURE 9.10   I/Q modulator: (*a*) simplified block diagram; (*b*) circuit realization.

In-phase (I)/quadrature-phase (Q) modulators are extensively used in communication systems for QPSK modulation. As shown in Fig. 9.10, the modulator is comprised of two double-balanced mixers. The mixers are fed at the LO ports by a carrier phase-shifted through a 3-dB 90° hybrid coupler. The carrier signal thus has a relative phase of 0° to one mixer and 90° to the other mixer. Modulation signals are fed externally in phase quadrature to the IF ports of the two mixers. The output modulated signals from the two mixers are combined through a two-way 3-dB in-phase power divider/combiner.

The 8-PSK consists of eight (2^3) phase states and a theoretical bandwidth efficiency of 3 bps/Hz. It transmits eight phases of 0°, 45°, 90°, 135°, 180°, 225°, 270°, and 315°. The 16-PSK transmits 16 phases. However, it is not used very much due to the small phase separation, which is difficult to maintain accurately. Instead, a modulation having both PSK and AM has evolved, called quadrature amplitude modulation (QAM). Figure 9.11 shows the output signal diagrams for 8-PSK, 16-PSK, and 16-QAM for comparison. Higher levels of QAM (64-, 256-, 1024-QAM)

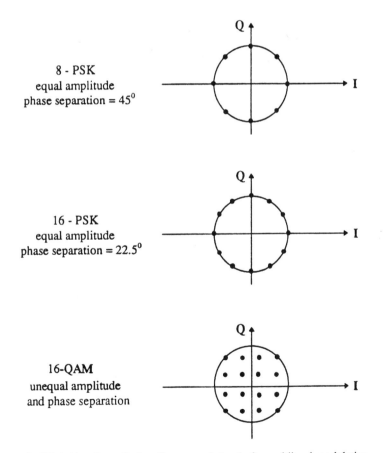

FIGURE 9.11 Constellation diagrams of signals for multilevel modulation.

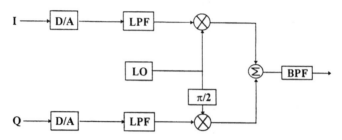

FIGURE 9.12   Quadrature amplitude modulator.

can be used for higher bandwidth efficiency. Figure 9.12 shows the typical QAM modulator block diagram. Two bit streams (I and Q) are provided from the pulse amplitude modulation process.

Some variations of QPSK are also in use. Offset-keyed or staggered quadriphase shift keying (OQPSK or SQPSK) modulation is used with only 90° phase transitions occurring in the modulator output signals. A $\frac{1}{4}\pi$-shifted, differentially encoded quadrature phase shift keying ($\frac{1}{4}\pi$-DQPSK) has been used for the U.S. and Japanese digital cellular time division multiple access (TDMA) radio standard; it has high power efficiency and spectral efficiency. In power-efficient, nonlinearly amplified (NLA) environments, where fully saturated class C amplifiers are used, the instantaneous 180° phase shift of conventional QPSK systems leads to a significant spectral regrowth and thus a low spectral efficiency. The OQPSK has 0° and ±90° phase transitions instead of 0°, ±90°, and ±180° for conventional QPSK. The compromise between conventional QPSK and OQPSK is $\frac{1}{4}\pi$-DQPSK, with 0°, ±45°, and ±135° phase transitions.

9.5   BIT ERROR RATE AND BANDWIDTH EFFICIENCY

A binary digital modulation system transmits a stream of data with binary symbols 0 and 1. The baseband information could be voice, music, fax, computer, or telemetry data. For analog information such as voice and music, an analog-to-digital (A/D) converter is used to convert the analog information into a digital form. The receiver will recover the data stream information.

In the ideal case, a receiver will recover the same binary digital stream that is transmitted, but the presence of noise in a communication system (e.g., transmitter, propagation, receiver) introduces the probability of errors that will be generated in the detection process. The likelihood that a bit is received incorrectly is called the bit error rate or the probability of error, defined as

$$BER = \frac{\text{false bits}}{\text{received bits}} \qquad (9.10)$$

Example 9.2 A communication system transmits data at a rate of 2.048 Mb/sec. Two false bits are generated in each second. What is the BER?

Solution

$$\text{BER} = \frac{2 \text{ b/sec}}{2.048 \text{ Mb/sec}} = 9.76 \times 10^{-7} \qquad \blacksquare$$

The BER can be reduced if the system's SNR is increased. Figure 9.13 shows the BERs as a function of SNR for various modulation levels [4]. It can be seen that the BER drops rapidly as the SNR increases. The higher levels of modulation give better bandwidth efficiency but would require higher values of SNR to achieve a given BER. This is a trade-off between bandwidth efficiency and signal (carrier) power. The desired signal power must exceed the combined noise and interference power by an amount specified by the SNR ratio. The lower the SNR, the higher the BER and the more difficult it is to reconstruct the desired data information.

Coherent communication systems can improve the BER. The LO in the receiver is in synchronism with the incoming carrier. Synchronism means that both frequency and phase are identical. This can be accomplished in two ways: transmitting a pilot carrier or using a carrier-recovery circuit. The transmitting carrier used as a reference will pass through the same propagation delays as the modulated signal and will arrive at the receiver with the same phase and frequency. The carrier can be used to phase-lock the receiver's LO. Figure 9.14 shows the improvement in BER for coherent systems as compared to noncoherent systems [5].

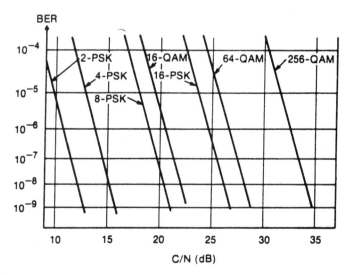

FIGURE 9.13 Bit error rates vs. carrier-to-noise ratio for different modulation schemes. (From reference [4], with permission from IEEE.)

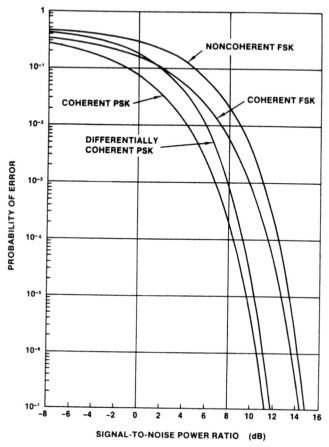

FIGURE 9.14 Comparison between coherent and noncoherent systems [5].

In binary digital modulation systems, if the system transmits 1 bit during each bit period, the system has a bandwidth efficiency of 1 bps/Hz. A bandwidth of 30 kHz can transmit $(30\,\text{kHz})(1\,\text{bps/Hz}) = 30\,\text{Kbps}$ data rate. For the n-PSK and n-QAM modulation, the total number of states (or phases) is given by

$$n = 2^M \tag{9.11}$$

The theoretical bandwidth efficiency (η) is equal to M bps/Hz. The bandwidth efficiencies for BPSK, QPSK, 8-PSK, and 16-PSK are 1, 2, 3, and 4 bps/Hz, respectively. In other words, one can transmit 2, 4, 8, and 16 phases per second for these different modulation levels. In practice, when the number of states is increased, the separation between two neighboring states becomes smaller. This will cause uncertainty and increase the BER. Nonideal filtering characteristics also limit the

TABLE 9.1 Modulation Bandwidth Efficiencies

M	Modulation	Theoretical Bandwidth Efficiencies (bps/Hz)	Actual Bandwidth Efficiencies (bps/Hz)
1	BPSK	1	1
2	QPSK	2	2
3	8-PSK	3	2.5
4	16-PSK/16-QAM	4	3
6	64-QAM	6	4.5
8	256-QAM	8	6

bandwidth efficiency [2]. Therefore, the actual bandwidth efficiency becomes smaller, given by

$$\eta = 0.75M \tag{9.12}$$

Equation (9.12) is normally used for high-level or multilevel modulation with $M \geq 4$. Table 9.1 summarizes the bandwidth efficiencies for different modulation schemes.

9.6 SAMPLING AND PULSE CODE MODULATION

For a continuous signal (analog signal), the signal will be sampled and pulse encoded before the digital modulation takes place. If the discrete sampling points have sufficiently close spacing, a smooth curve can be drawn through them. It can be therefore said that the continuous curve is adequately described by the sampling points alone. One needs to transmit the sampling points instead of the entire continuous signal.

Figure 9.15 illustrates the sampling process and the results after sampling. The samples are then quantized, encoded, and modulated, as shown in the block diagram in Fig. 9.16. This process is called pulse code modulation (PCM). The advantages of this process are many:

1. The transmitted power can be concentrated into short bursts rather than delivered in CW. Usually, the pulses are quite short compared to the time between them, so the source is "off" most of the time. Peak power higher than the CW power can thus be transmitted.
2. The time intervals between pulses can be filled with sample values from other messages, thereby permitting the transmission of many messages on one communication system. This time-sharing transmission is called time division multiplexing.
3. The message is represented by a coded group of digital pulses. The effects of random noise can be virtually eliminated.

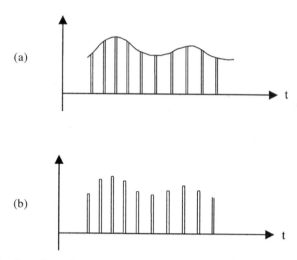

(a)

(b)

FIGURE 9.15 Sampling of a continuous signal: (*a*) continuous signal and sampling points; (*b*) sampled signal.

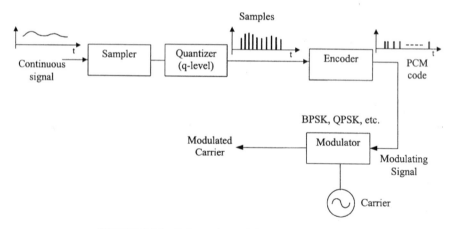

FIGURE 9.16 Pulse code modulation block diagram.

As shown in Fig. 9.16, the continuous signal is first sampled. The sample values are then rounded off to the nearest predetermined discrete value (quantum value). The encoder then converts the quantized samples into appropriate code groups, one group for each sample, and generates the corresponding digital pulses forming the baseband. This is basically an A/D converter.

The quantization is done depending on the number of pulses used. If N is the number of pulses used for each sample, the quantized levels q for a binary system are given by

$$q = 2^N \tag{9.13}$$

Figure 9.17 shows an example of sampling and encoding with $N = 3$. The more quantized levels used, the more accurate the sample data can be represented. However, it will require more bits (pulses) per sample transmission. Table 9.2 shows the number of bits per sample versus the number of quantizing steps. The PCM codes will be used as the modulating signal (information) to a digital modulator.

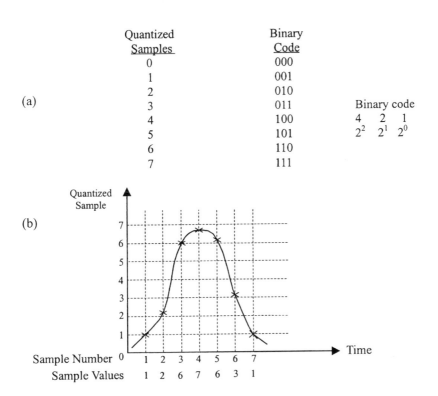

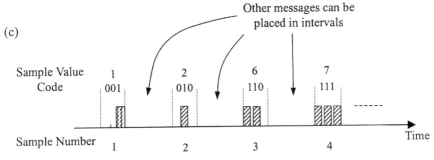

FIGURE 9.17   Sampling and PCM encoding for $N = 3$: (*a*) quantized samples and binary codes; (*b*) sampling and quantization; (*c*) PCM codes.

TABLE 9.2   Quantizing Steps

Number of Bits, N	Number of Quantizing Steps, q
3	8
4	16
5	32
6	64
7	128
8	256
9	512
10	1024
11	2048
12	4096

PROBLEMS

9.1 In an AM broadcast system, the total power transmitted is 1000 W. Calculate the transmitted power at the carrier frequency and at the upper and lower sidebands for an 80% modulation.

9.2 The power content for the carrier in an AM modulation is 1 kW. Determine the power content of each of the sidebands and the total transmitted power when the carrier is modulated at 75%.

9.3 Explain how the balanced modulator shown in Fig. P9.3 works.

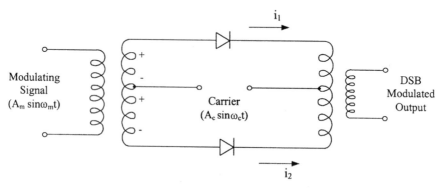

FIGURE P9.3

9.4 A 107.6-MHz carrier is frequency modulated by a 5-kHz sine wave. The frequency deviation is 50 kHz.

(a) Determine the carrier swing of the FM signal.

(b) Determine the highest and lowest frequencies attained by the modulated signal.

(c) What is the modulation index?

9.5 A QPSK communication link has a BER of 10^{-6}. The system data rate is 200 Mb/sec. Calculate the bandwidth requirement and the number of false bits generated per second. What is the bandwidth requirement if one uses 64-QAM modulation for the same data rate?

9.6 A QPSK mobile communication system has a maximum range of 20 km for a receiver output SNR of 10 dB. What is the SNR if the system is operated at a distance of 10 km? At the maximum range, if the system has 10 false bits per second for the transmission of 1 Mb/sec, what is the BER of this system?

9.7 For the signal shown in Fig. P9.7, a sample is generated every 1 msec, and four pulses are used for each sample (i.e., $N = 4$). Determine the quantized samples and binary codes used. What are the sampled values at each sampling time?

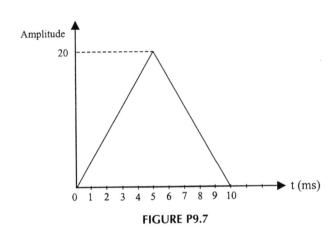

FIGURE P9.7

REFERENCES

1. K. Chang, *Microwave Solid-State Circuits and Applications*, John Wiley & Sons, New York, 1994.
2. R. G. Winch, *Telecommunication Transmission Systems*, McGraw-Hill, New York, 1993.
3. A. Grote and K. Chang, "60-GHz Integrated-Circuit High Data Rate Quadriphase Shift Keying Exciter and Modulator," *IEEE Trans. Microwave Theory Tech.*, Vol. MTT-32, No. 12, pp. 1663–1667, 1984.
4. T. Noguchi, Y. Daido, and J. A. Nossek, "Modulation Techniques for Microwave Digital Radio," *IEEE Commun. Mag.*, Vol. 24, No. 10, pp. 21–30, 1986.
5. E. A. Wolff and R. Kaul, *Microwave Engineering and Systems Applications*, John Wiley & Sons, New York, 1988.

![black bar](under CHAPTER TEN)

CHAPTER TEN

Multiple-Access Techniques

10.1 INTRODUCTION

Three commonly used techniques for accommodating multiple users in wireless communications are frequency division multiple access (FDMA), time division multiple access (TDMA), and code division multiple access (CDMA). Frequency division multiple access and TDMA are old technologies and have been used for quite a while. Code division multiple access is the emerging technology for many new cellular phone systems. This chapter will briefly discuss these techniques.

10.2 FREQUENCY DIVISION MULTIPLE ACCESS AND FREQUENCY DIVISION MULTIPLEXING

For the FDMA and frequency division multiplexing (FDM) systems, the available frequency band is split into a specific number of channels, and the bandwidth of each channel depends on the type of information to be transmitted. To transmit a number of channels over the same system, the signals must be kept apart so that they do not interfere with each other.

Figure 10.1 shows an example of the FDM transmitter system with simultaneous transmission of 10 signals from 10 users. Each signal contains video information from 0 to 6 MHz with a guard band of 4 MHz. A double side band (DSB) modulator is used. The guard band is placed between two adjacent signals to avoid interference. A multiplexer is used to combine the signals, and the combined signals are then upconverted and amplified.

In the receiver, the signals are separated by a multiplexer that consists of many filters. The information is recovered after the demodulator. Figure 10.2 shows a receiver block diagram. The advantage of FDMA is that no network timing

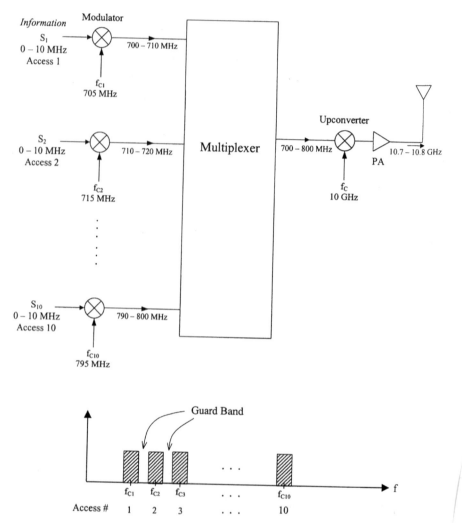

FIGURE 10.1 FDM system and frequency spectrums.

is required, and the major disadvantages include required power control, a wide frequency band, and interference caused by intermodulation and sideband distortion.

10.3 TIME DIVISION MULTIPLE ACCESS AND TIME DIVISION MULTIPLEXING

A TDMA or time division multiplexing (TDM) system uses a single frequency ba to simultaneously transmit many signals (channels) in allocated time slots. Th different channels time-share the same frequency band without interfering with e

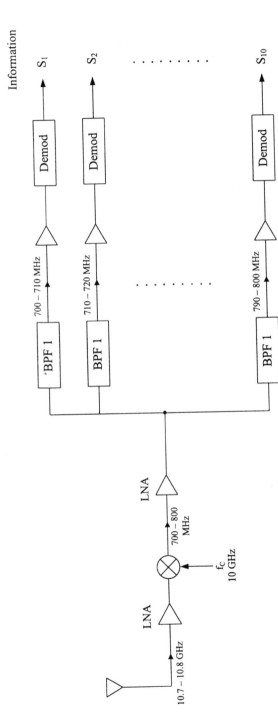

FIGURE 10.2 Receiver block diagram for an FDM system.

other. The advantages of TDMA as compared to FDMA are the requirement of a narrower frequency bandwidth, invulnerability to interchannel crosstalk and imperfect channel filtering, no power control required, and high efficiency. The disadvantage is the requirement of network timing.

Figure 10.3*a* shows a block diagram of a TDMA transmitting system. The samples are interleaved, and the composite signal consists of all of the interleaved pulses. A commutator or switch circuit is normally used to accomplish the data interleaving. Figure 10.3*b* shows an example of TDM of two signals. Figure 10.3*c* shows the data slot allocation for N signals. Each data slot could consist of a group

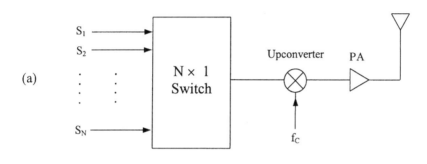

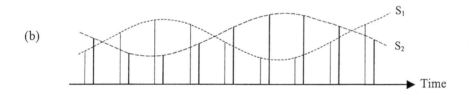

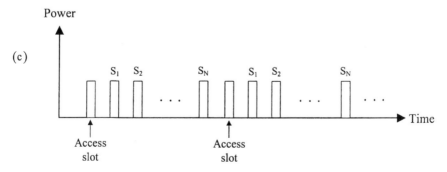

FIGURE 10.3 TDMA or TDM system: (*a*) a transmitter; (*b*) TDM of two signals; (*c*) data slot allocation for N signals.

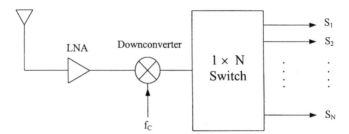

FIGURE 10.4   TDMA receiver.

of PCM codes. All samples are transmitted sequentially. At the receiver, the composite signal is demultiplexed by using a $1 \times N$ switch or commutator (Fig. 10.4).

10.4   SPREAD SPECTRUM AND CODE DIVISION MULTIPLE ACCESS

Spread spectrum (SS) is broadly defined as a technique by which the transmitted signal bandwidth is much greater than the baseband information signal bandwidth. The technique was initially developed by the military since it provided the desirable advantage of having a low probability of detection and thus made for secure communications. Because today's cellular and mobile communication systems suffer from severe spectrum congestion, especially in urban areas, spread spectrum techniques are used to increase system capacity in order to relieve congestion.

Spread spectrum has the following features and advantages:

1. It improves the interference rejection.
2. Because each user needs a special code to get access to the data stream, it has applications for secure communications and code division multiple access.
3. It has good antijamming capability.
4. The capacity and spectral efficiency can be increased by the use of spread spectrum techniques. Many users can use the same frequency band with different codes.
5. It has a nice feature of graceful degradation as the number of users increases.
6. Low-cost IC components can be used for implementation.

In the implementation of the spread spectrum technique, a modulated signal is modulated (spread) a second time to generate an expanded-bandwidth wide-band signal that does not interfere with other signals. The second modulation can be accomplished by one of the following methods [1]:

1. Direct-sequence spread spectrum (DSSS)
2. Frequency-hopping spread spectrum (FHSS)

3. Time hopping
4. Chirp

Figure 10.5*a* shows an example of a transmitter for the DSSS system [2]. The digital binary information is first used to modulate the IF carrier. The modulated signal is

$$s(t) = A \cos(\omega_{IF}t + \phi) \tag{10.1}$$

where $\phi = 0°$ and $\phi = 180°$ for a BPSK modulator when the data are 1 and 0. The modulated IF carrier is modulated again by a spreading signal function $g(t)$, where $g(t)$ could be a pseudonoise (PN) signal or a code signal. Each user is assigned a special code. The output signal is equal to

$$v(t) = g(t)s(t) = Ag(t) \cos(\omega_{IF}t + \phi) \tag{10.2}$$

(a)

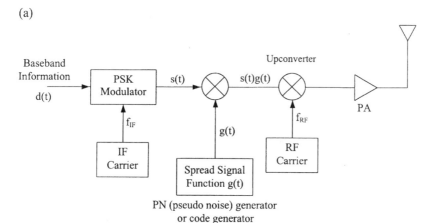

(b)

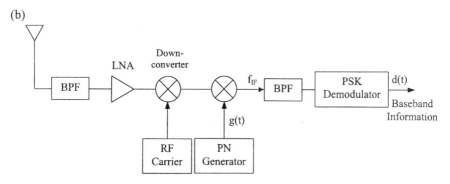

FIGURE 10.5   Direct-sequence spread spectrum system: (*a*) transmitter; (*b*) receiver.

This signal is upconverted by an RF carrier obtained from a phase-locked source or a frequency synthesizer. The signal is finally amplified and transmitted through an antenna. At the receiver end (Fig. 10.5b), the intended user will have a synchronized $g(t)$ that despreads the received signal and has the same PN sequence as that of the corresponding transmitter. The despreading or decoded signal is PSK demodulated to recover the information. Although the BPSK is assumed here for the first modulation, other digital modulation techniques described in Chapter 9 can be used.

The FHSS is similar to the direct-sequence spread spectrum system. The difference is that the PN sequence generator is used to control a frequency synthesizer to hop to one of the many available frequencies chosen by the PN sequence generator. As shown in Fig. 10.6, the output frequencies from the synthesizer hop pseudorandomly over a frequency range covering $f_1, f_2, \ldots, f_N$. Since N could be several thousand or more, the spectrum is spread over a wide frequency range. These output frequencies are the RF carrier frequencies coupled to the upconverter. The system is called a fast frequency hopping spread spectrum (FFHSS) system if the hopping rate is higher than the data bit rate. If the hopping rate is slower than the data rate, the system is called a slow frequency hopping spread spectrum (SFHSS) system. In the receiver, a PN generator with the same sequence (code) is used to generate the same frequency hopping sequence. These frequencies are used to downconvert the received signal. The IF signal is then demodulated to recover the data.

An important component in the spread spectrum (SS)-CDMA system is the PN sequence (or PN code) generator. The major functions of the PN code generator are as follows:

1. Spread the bandwidth of the modulated signal to the larger transmission bandwidth.
2. Distinguish between the different user signals utilizing the same transmission bandwidth in a multiple-access scheme.

The PN code is the "key" of each user to access his or her intended signal in the receiver. Two commonly used sequences are the maximal-length sequences and Gold sequences. The maximal-length sequences (m-sequences) use cascaded flip-flops to generate the random codes. As shown in Fig. 10.7, each flip-flop can generate a logic output of 1 or 0. If N is the total number of flip-flops, the sequence length L in bits is given by

$$L = 2^N - 1 \tag{10.3}$$

As examples, if $N = 3$, $L = 2^3 - 1 = 7$. If $N = 15$, $L = 2^{15} - 1 = 32{,}767$. The subtraction of 1 in Eq. (10.3) is to exclude the code with all zeroes. Here, L represents the maximum number of users with different codes.

Gold sequences were invented by R. Gold in 1967 [3]. Gold sequences are generated by combining two m-sequences clocked by the same chip-clock, as shown

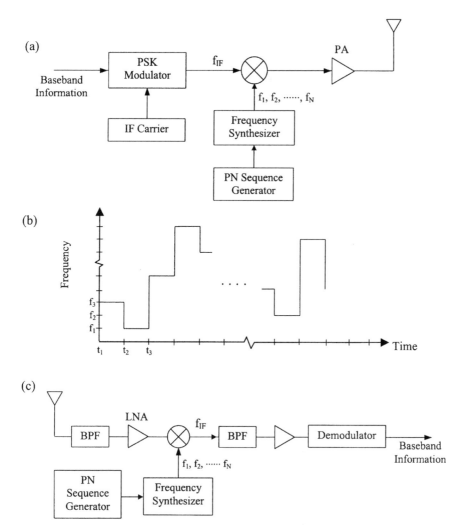

FIGURE 10.6   Frequency hopping spread spectrum system: (*a*) transmitter block diagram; (*b*) frequency hopping output; (*c*) receiver block diagram.

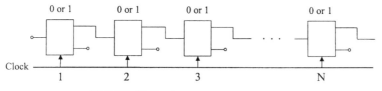

FIGURE 10.7   An m-sequence generator.

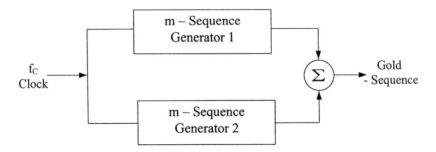

FIGURE 10.8   Gold sequence generator.

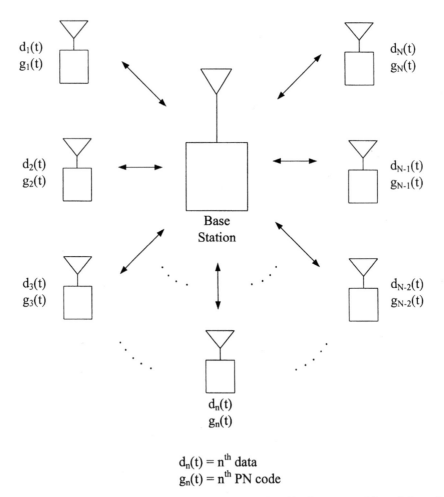

$$d_n(t) = n^{th} \text{ data}$$
$$g_n(t) = n^{th} \text{ PN code}$$

FIGURE 10.9   Many users share the same frequency band in the same mobile cellular cell using CDMA techniques.

in Fig. 10.8. The Gold sequences offer good cross-correlation between the single sequences.

In summary, SS-CDMA received widespread interest because it allows many users to occupy the same frequency band without causing interference. In military applications, it offers secure communications and immunity to jamming. In wireless mobile communications, it allows many users to simultaneously occupy the same frequency. Each user is assigned a unique code. All signals from all users are received by each user, but each receiver is designed to listen to and recognize only one specific sequence. Figure 10.9 shows many users communicating simultaneously with the base station operating at the same frequency band. Compared to TDMA systems, CDMA has the following advantages: (1) It is relatively easy to add new users. (2) It has the potential for higher capacity. (3) The system is more tolerant to multipath fading and more immune to interference. (4) Network synchronization is not required. Because of these advantages, many new communications will use CDMA techniques.

REFERENCES

1. G. R. Cooper and C. D. McGillem, *Modern Communications and Spread Spectrum*, McGraw-Hill, New York, 1986.

2. K. Feher, *Wireless Digital Communications*, Prentice-Hall, Upper Saddle River, NJ, 1995.

3. R. Gold, "Optimal Binary Sequences for Spread Spectrum Multiplexing," *IEEE Trans. Inform. Theory*, Vol. IT-13, pp. 619–621, Oct. 1967.

CHAPTER ELEVEN

Other Wireless Systems

The two major applications of RF and microwave technologies are in communications and radar/sensor systems. Radar and communication systems have been discussed in Chapters 7 and 8, respectively. There are many other applications such as navigation and global positioning systems, automobile and highway applications, direct broadcast systems, remote sensing, RF identification, surveillance systems, industrial sensors, heating, environmental, and medical applications. Some of these systems will be discussed briefly in this chapter. It should be emphasized that although the applications are different, the general building blocks for various systems are quite similar.

11.1 RADIO NAVIGATION AND GLOBAL POSITIONING SYSTEMS

Radio navigation is a method of determining position by measuring the travel time of an electromagnetic (EM) wave as it moves from transmitter to receiver. There are more than 100 different types of radio navigation systems in the United States. They can be classified into two major kinds: active radio navigation and passive radio navigation, shown in Figs. 11.1 and 11.2.

Figure 11.1 shows an example of an active radio navigation system. An airplane transmits a series of precisely timed pulses with a carrier frequency f_1. The fixed station with known location consists of a transponder that receives the signal and rebroadcasts it with a different frequency f_2. By comparing the transmitting and receiving pulses, the travel time of the EM wave is established. The distance between the aircraft and the station is

$$d = c(\tfrac{1}{2}t_R) \tag{11.1}$$

where t_R is the round-trip travel time and c is the speed of light.

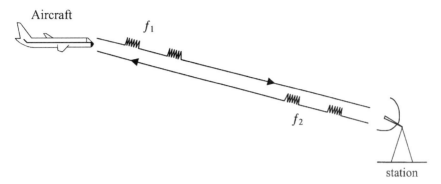

FIGURE 11.1   Active radio navigation system.

In a passive radio navigation system, the station transmits a series of precisely timed pulses. The aircraft receiver picks up the pulses and measures the travel time. The distance is calculated by

$$d = ct_R \tag{11.2}$$

where t_R is the one-way travel time.

The uncertainty in distance depends on the time measurement error given in the following:

$$\Delta d = c \, \Delta t_R \tag{11.3}$$

If the time measurement has an error of 10^{-6} s, the distance uncertainty is about 300 m.

To locate the user position coordinates, three unknowns need to be solved: altitude, latitude, and longitude. Measurements to three stations with known locations will establish three equations to solve the three unknowns. Several typical radio navigation systems are shown in Table 11.1 for comparison. The Omega

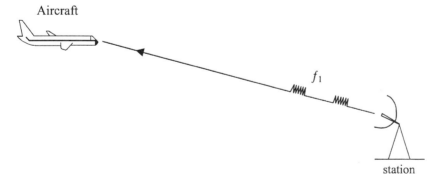

FIGURE 11.2   Passive radio navigation system.

TABLE 11.1 Comparison of Radio Navigation Systems

System	Position Accuracy, m[a]	Velocity Accuracy, m/sec	Range of Operation	Comments
Global Positioning Systems, GPS	16 (SEP)	≥ 0.1 (rms per axis)[b]	Worldwide	24-h all-weather coverage; specified position accuracy available to authorized users
Long-Range Navigation, Loran C[c]	180 (CEP)	No velocity data	U.S. coast and continental, selected overseas areas	Localized coverage; limited by skywave interference
Omega	2200 (CEP)	No velocity data	Worldwide	24-h coverage; subject to VLF propagation anomalies
Standard inertial navigation systems, Std INS[d]	≤ 1500 after 1st hour (CEP)	0.8 after 2 h (rms per axis)	Worldwide	24-h all-weather coverage; degraded performance in polar areas
Tactical Air Navigation, Tacan[c]	400 (CEP)	No velocity data	Line of sight (present air routes)	Position accuracy is degraded mainly by azimuth uncertainty, which is typically on the order of 1.0°
Transit[c]	200 (CEP)	No velocity data	Worldwide	90-min interval between position fixes suits slow vehicles (better accuracy available with dual-frequency measurements)

[a]SEP, CEP = spherical and circular probable error (linear probable error in three and two dimensions).
[b]Dependent on integration concept and platform dynamics.
[c]Federal Radionavigation Plan, December 1984.
[d]SNU-84-1 *Specification for USAF Standard Form Fit and Function (F³) Medium Accuracy Inertial Navigation Set/Unit*, October 1984.
Source: From reference [1], with permission from IEEE.

system uses very low frequency. The eight Omega transmitters dispersed around the globe are located in Norway, Liberia, Hawaii, North Dakota, Diego Garcia, Argentina, Australia, and Japan. The transmitters are phase locked and synchronized, and precise atomic clocks at each site help to maintain the accuracy. The use of low frequency can achieve wave ducting around the earth in which the EM waves bounce back and forth between the earth and ionosphere. This makes it possible to use only eight transmitters to cover the globe. However, the long wavelength at low frequency provides rather inaccurate navigation because the carrier cannot be modulated with useful information. The use of high-frequency carrier waves, on the other hand, provides better resolution and accuracy. But each transmitter can cover only a small local area due to the line-of-sight propagation as the waves punch through the earth's ionosphere. To overcome these problems, space-based satellite systems emerged. The space-based systems have the advantages of better coverage, an unobstructed view of the ground, and the use of higher frequency for better accuracy and resolution.

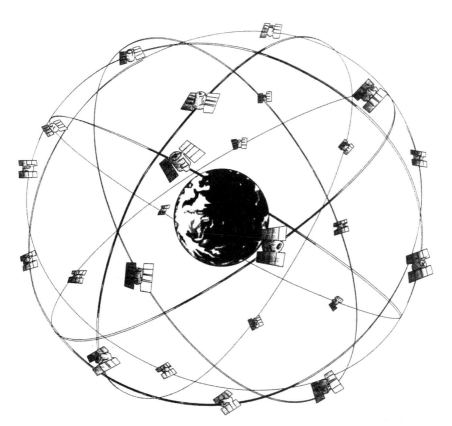

FIGURE 11.3 Navstar global positioning system satellite. (From reference [1], with permission from IEEE.)

The 24 Navstar global positioning satellites have been launched into 10,898 nautical mile orbits (approximately 20,200 km, 1 nautical mile = 1.8532 km) in six orbital planes. Four satellites are located in each of six planes at 55° to the plane of the earth's equator, as shown in Fig. 11.3. Each satellite continuously transmits pseudorandom codes at two frequencies (1227.6 and 1575.42 MHz) with accurately synchronized time signals and data about its own position. Each satellite covers about 42% of the earth.

The rubidium atomic clock on board weighs 15 lb, consumes 40 W of power, and has a timing stability of 0.2 parts per billion [2]. As shown in Fig. 11.4, the timing signal from three satellites would be sufficient to nail down the receiver's three position coordinates (altitude, latitude, and longitude) if the Navstar receiver is synchronized with the atomic clock on board the satellites. However, synchronization of the receiver's clock is in general impractical. An extra timing signal from the fourth satellite is used to solve the receiver's clock error. The user's clock determines a pseudorange R' to each satellite by noting the arrival time of the signal. Each of the four R' distances includes an unknown error due to the inaccuracy of the user's inexpensive clock. In this case, there are four unknowns: altitude, latitude, longitude, and clock error. It requires four measurements and four equations to solve these four unknowns.

Figure 11.5 shows the known coordinates of four satellites and the unknown coordinates of the aircraft, for example. The unknown x, y, z represent the longitude,

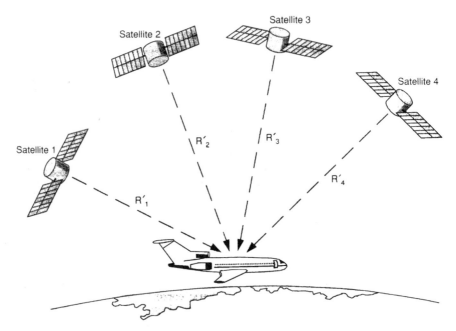

FIGURE 11.4   Determination of the aircraft's position. (From reference [1], with permission from IEEE.)

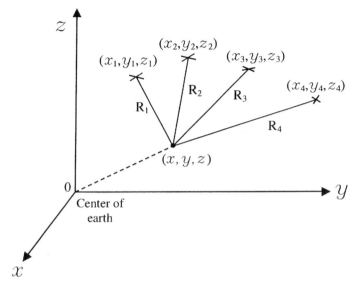

FIGURE 11.5 Coordinates for four satellites and a user.

latitude, and altitude, respectively, measured from the center of the earth. The term ε represents the receiver clock error. Four equations can be set up as follows:

$$[(x_1 - x)^2 + (y_1 - y)^2 + (z_1 - z)^2]^{1/2} = c(\Delta t_1 - \varepsilon) = R_1 \qquad (11.4a)$$

$$[(x_2 - x)^2 + (y_2 - y)^2 + (z_2 - z)^2]^{1/2} = c(\Delta t_2 - \varepsilon) = R_2 \qquad (11.4b)$$

$$[(x_3 - x)^2 + (y_3 - y)^2 + (z_3 - z)^2]^{1/2} = c(\Delta t_3 - \varepsilon) = R_3 \qquad (11.4c)$$

$$[(x_4 - x)^2 + (y_4 - y)^2 + (z_4 - z)^2]^{1/2} = c(\Delta t_4 - \varepsilon) = R_4 \qquad (11.4d)$$

where R_1, R_2, R_3, and R_4 are the exact ranges. The pseudoranges are $R'_1 = c\,\Delta t_1$, $R'_2 = c\,\Delta t_2$, $R'_3 = c\,\Delta t_3$, and $R'_4 = c\,\Delta t_4$. The time required for the signal traveling from the satellite to the receiver is Δt.

We have four unknowns (x, y, z, and ε) and four equations. Solving Eqs. (11.4a)–(11.4d) results in the user position information (x, y, z) and ε. Accuracies of 50–100 ft can be accomplished for a commercial user and better than 10 ft for a military user.

11.2 MOTOR VEHICLE AND HIGHWAY APPLICATIONS

One of the biggest and most exciting applications for RF and microwaves is in automobile and highway systems [3–6]. Table 11.2 summarizes these applications. Many of these are collision warning and avoidance systems, blind-spot radar, near-obstacle detectors, autonomous intelligent cruise control, radar speed sensors, optimum speed data, current traffic and parking information, best route information,

TABLE 11.2 Microwave Applications on Motor Vehicles and Highways

I. Motor vehicle applications
   Auto navigation aids and global positioning systems
   Collision warning radar
   Automotive telecommunications
   Speed sensing
   Antitheft radar or sensor
   Blind spot detection
   Vehicle identification
   Adaptive cruise control
   Automatic headway control
   Airbag arming
II. Highway and traffic management applications
   Highway traffic controls
   Highway traffic monitoring
   Toll-tag readers
   Vehicle detection
   Truck position tracking
   Intelligent highways
   Road guidance and communication
   Penetration radar for pavement
   Buried-object sensors
   Structure inspection

and the Intelligent Vehicle and Highway System (IVHS). One example of highway applications is automatic toll collection. Automatic toll collection uses Automatic Vehicle Identification (AVI) technology, which provides the ability to uniquely identify a vehicle passing through the detection area. As the vehicle passes through the toll station, the toll is deducted electronically from the driver's account. Generally, a tag or transponder located in the vehicle will answer an RF signal from a roadside reader by sending a response that is encoded with specific information about the vehicle or driver. This system is being used to reduce delay time and improve traffic flow.

A huge transportation application is IVHS. The IVHS systems are divided into five major areas. Advanced Traveler Information Systems (ATIS) will give navigation information, including how to find services and taking into account current weather and traffic information. Advanced Traffic Management Systems (ATMS) will offer real-time adjustment of traffic control systems, including variable signs to communicate with motorists. Advanced Vehicle Control Systems (AVCS) will identify upcoming obstacles, adjacent vehicles, and so on, to assist in preventing collisions. This is intended to evolve into completely automated highways. Commercial Vehicle Operations (CVO) will offer navigation information tailored to commercial and emergency vehicle needs in order to improve efficiency and safety. Finally, Advanced Public Transit Systems (APTS) will address the mass transit needs of the public. All of these areas rely heavily on microwave data communications that can be broken down into four categories: intravehicle, vehicle to vehicle, vehicle to infrastructure, and infrastructure to infrastructure.

Since the maximum speed in Europe is 130 km/hr, the anticollision radars being developed typically require a maximum target range of around 100 m. Detecting an object at this distance gives nearly 3 s warning so that action can be taken. Anticollision systems should prove to be most beneficial in low-visibility situations, such as fog and rain. Systems operating all over the frequency spectrum are being developed, although the 76–77 GHz band has been very popular for automotive anticollision radars. Pulsed and FM CW systems are in development that would monitor distance, speed, and acceleration of approaching vehicles. European standards allow a 100-MHz bandwidth for FM CW systems and a 500-MHz bandwidth for pulsed systems. Recommended antenna gain is 30–35 dB with an allowed power of 16–20 dBm. Fairly narrow beamwidths (2.5° azimuth, 3.5° elevation) are necessary for anticollision radar so that reflections are received only from objects in front of or behind the vehicle and not from bridges or objects in other lanes. Because of this, higher frequencies are desirable to help keep antenna size small and therefore inconspicuous. Multipath reflections cause these systems to need 6–8 dB higher power than one would expect working in a single-path environment. Figure 11.6 shows an example block diagram for a forward-looking automotive radar (FLAR) [7].

A nonstop tolling system named Pricing and Monitoring Electronically of Automobiles (PAMELA) is currently undergoing testing in the United Kingdom. It is a 5.8-GHz system that utilizes communication between a roadside beacon mounted on an overhead structure and a passive transponder in the vehicle. The roadside beacon utilizes a circularly polarized 4 × 4 element patch antenna array with a 17-dB gain and a 20° beamwidth. The vehicle transponder uses a 120° beamwidth. This sytem has been tested at speeds up to 50 km/hr with good results. The system is intended to function with speeds up to 160 km/hr.

Automatic toll debiting systems have been allocated to the 5.795–5.805- and 5.805–5.815-GHz bands in Europe. This allows companies either two 10-MHz channels or four 5-MHz channels. Recommended antenna gain is 10–15 dB with an allowed power of 3 dBm. Telepass is such an automatic toll debiting system installed along the Milan–Naples motorway in Italy. Communication is over a 5.72-GHz link. A SMART card is inserted into the vehicle transponder for prepayment or direct deduction from your bank account. Vehicles slow to 50 km/hr for communication, then resume speed. If communication cannot be achieved, the driver is directed to another lane for conventional payment.

Short Range Microwave Links for European Roads (SMILER) is another system for infrastructure to vehicle communications. Transmission occurs at 61 GHz between a roadside beacon and a unit on top of the vehicle. Currently horn antennas are being used on both ends of the link, and the unit is external to the vehicle to reduce attenuation. The system has been tested at speeds up to 145 km/hr with

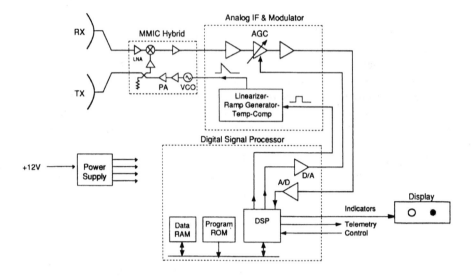

Parameters	Requirements
Range	2-100 meters
Range Resolution	0.5 meters
Range Rate Resolution	1.5 km/hr
Antenna Beamwidth	3 by 3 degrees, Linear
Antenna Sidelobes	< -30 dB
Modulation Type	FMCW
Modulation Bandwidth	350 MHz
Transmit Power	> 10 mW
S/N Ratio	10 dB
Display	Laptop Computer via RS232
DC Power Supply	12 volts

FIGURE 11.6 Block diagram and specifications of a W-band forward-looking automotive radar system. (From reference [7], with permission from IEEE.)

single-lane discrimination. SMILER logs the speed of the vehicle as well as transmitting information to it.

V-band communication chips developed for defense programs may see direct use in automotive communications either from car to car or from car to roadside. The 63–64-GHz band has been allocated for European automobile transmissions. An MMIC-based, 60-GHz receiver front end was constructed utilizing existing chips.

Navigation systems will likely employ different sources for static and dynamic information. Information such as road maps, gas stations, and hotels/motels can be displayed in the vehicle on color CRTs. Dynamic information such as present location, traffic conditions, and road updates would likely come from roadside communication links or GPS satellites.

11.3   DIRECT BROADCAST SATELLITE SYSTEMS

The direct broadcast satellite (DBS) systems offer a powerful alternative to cable television. The system usually consists of a dish antenna, a feed horn antenna, an MMIC downconverter, and a cable to connect the output of the downconverter to the home receiver/decoder and TV set. For the C-band systems, the dish antenna is big with a diameter of 3 m. The X-band systems use smaller antennas with a diameter of about 3 ft. The new Ku-band system has a small 18-in. dish antenna. The RCA Ku-band digital satellite system (DirecTV) carries more than 150 television channels.

For all DBS systems, a key component is the front-end low-noise downconverter, which converts the high microwave signal to a lower microwave or UHF IF signal for low-loss transmission through the cable [8, 9]. The downconverter can be a MMIC GaAs chip with a typical block diagram shown in Fig. 11.7. Example specifications for a downconverter from ANADIGICS are shown in Table 11.3 [10]. The chip accepts an RF frequency ranging from 10.95 to 11.7 GHz. With an LO frequency of 10 GHz, the IF output frequency is from 950 to 1700 MHz. The system has a typical gain of 35 dB and a noise figure of 6 dB. The local oscillator phase noise is −70 dBc/Hz at 10 kHz offset from the carrier and −100 dBc/Hz at 100 kHz offset from the carrier.

The DBS system is on a fast-growth track. Throughout the United States, Europe, Asia, and the rest of the world, the number of DBS installations has rapidly increased. It could put a serious dent in the cable television business.

11.4   RF IDENTIFICATION SYSTEMS

Radio frequency identification (RFID) was first used in World War II to identify the friendly aircraft. Since then, the use has grown rapidly for a wide variety of applications in asset management, inventory control, security systems, access control, products tracking, assembly-line management, animal tracking, keyless

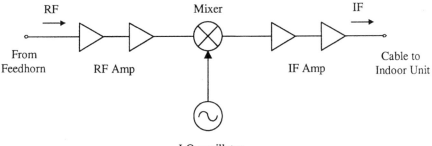

FIGURE 11.7   DBS downconverter block diagram.

TABLE 11.3   Specifications for an ANADIGICS Downconverter

Parameter	Minimum	Typical	Maximum	Units
Conversion gain				
$F_{RF} = 10.95$ GHz	32	35		dB
$F_{RF} = 11.7$ GHz	32	35		dB
SSB noise figure				
$F_{RF} = 10.95$ GHz		6.0	6.5	dB
$F_{RF} = 11.7$ GHz		6.0	6.5	dB
Gain flatness		±1.5	±2	dB
Gain ripple over any 27-MHz band		< 0.25		dB
LO–RF leakage		−25	−10	dBm
LO–IF leakage		−5	0	dBm
LO phase noise				
10 KHz offset		−70	−50	dBc/Hz
100 KHz offset		−100	−70	dBc/Hz
Temperature stability of LO		±1.5		MHz/°C
Image rejection	0	5		dB
Output power at 1 dB gain compression	0	−6		dBm
Output third-order IP	+10	+16		dBm
Power supply current				
I_{DD}	75	120	150	mA
I_{SS}	1	3.5	4	mA
Spurious output in any band			−60	dBm
Input VSWR with respect to 50 Ω over RF band		2 : 1		
Output VSWR with respect to 75 Ω over IF Band		1.5 : 1		

entry, automatic toll debiting, and various transportation uses. In fact, just about anything that needs to be identified could be a candidate for RFID. In most cases, the identification can be accomplished by bar-coded labels and optical readers commonly used in supermarkets or by magnetic identification systems used in libraries. The bar-coded and magnetic systems have the advantage of lower price tags as compared to RFID. However, RFID has applications where other less expensive approaches are ruled out due to harsh environments (where dust, dirt, snow, or smoke are present) or the requirement of precise alignment. The RFID is a noncontacting technique that has a range from a few inches to several hundred feet depending on the technologies used. It does not require a precise alignment between the tag and reader. Tags are generally reusable and can be programmed for different uses.

The RFID tags have been built at many different frequencies from 50 kHz to 10 GHz [11]. The most commonly used frequencies are 50–150 kHz, 260–470 MHz,

TABLE 11.4   Comparison of Systems in Different Frequencies

Parameter	VLF	HF	VHF	UHF	Microwave
Cost	Low	Low	Medium	Medium	High
Interference	Low	High	High	High	Low
Absorption	Low	Low	Medium	Medium	High
Reflection	None	Low	Medium	High	High
Data rate	Low	Medium	Medium	High	High

902–928 MHz, and 2450 MHz. Trade-offs of these frequencies are given in Table 11.4.

The RFID systems can be generally classified as coded or uncoded with examples shown in Fig. 11.8. In the uncoded system example, the reader transmits an interrogating signal and the tag's nonlinear device returns a second-harmonic signal. This system needs only the pass/fail decision without the necessity of the identification of the individual tag. For the coded systems, each tag is assigned an identification code and other information, and the returned signal is modulated to contain the coded information. The reader decodes the information and stores it in the data base. The reply signal may be a reflection or retransmission of the interrogating signal with added modulation, a harmonic of the interrogating signal with added modulation, or a converted output from a mixer with a different frequency (similar to the transponders described in Section 8.7 for satellite communications). The complexity depends on applications and system requirements. For example, the RFID system in air traffic control can be quite complex and the one used for antishoplifting very simple. The complex system normally requires a greater power supply, a sensitive receiver, and the reply signal at a different frequency than the interrogator.

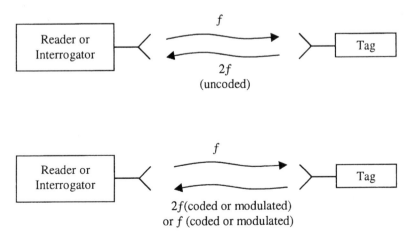

FIGURE 11.8   Simplified RFID systems.

The tags can be classified as active, driven (passive), and passive [12]. The active tag needs a battery; the driven and passive tags do not. The driven tags do need external power, but the power is obtained by rectifying the RF and microwave power from the interrogating signal or by using solar cells. The driven tags could use a diode detector to convert part of the interrogating signal into the DC power, which is used to operate the code generation, modulation, and other electronics.

The low-cost antitheft tags used for stores or libraries are uncoded passive tags. They are usually inexpensive diode frequency doublers that radiate a low-level second harmonic of the interrogating signal. Reception of the harmonic will alert the reader and trigger the alarm.

Numerous variations are possible depending on code complexity, power levels, range of operation, and antenna type. Four basic systems are shown in Fig. 11.9 [12]. The DC power level generated depends on the size of the tag antenna and the RF-to-DC conversion efficiency of the detector diode.

The passive and driven tags are usually operating for short-range applications. The battery-powered active tags can provide a much longer range with more complicated coding. If the size is not a limitation, larger batteries can be used to provide whatever capacity is required. The battery normally can last for several years of operation.

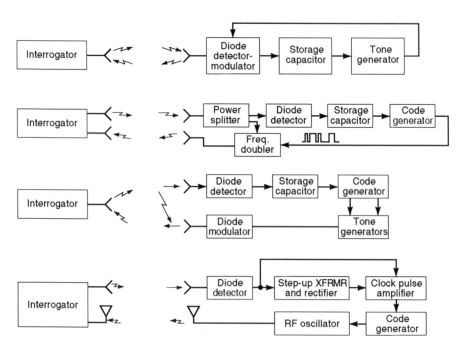

FIGURE 11.9 Four basic types of driven tags. (From reference [12] with permission from *RCA Review.*)

11.5   REMOTE SENSING SYSTEMS AND RADIOMETERS

Radiometry or microwave remote sensing is a technique that provides information about a target from the microwave portion of the blackbody radiation (noise). The radiometer normally is a passive, high-sensitivity (low-noise), narrow-band receiver that is designed to measure this noise power and determine its equivalent brightness temperature.

For an ideal blackbody, in the microwave region and at a temperature T, the noise and energy radiation is

$$P = kTB \qquad (11.5)$$

where k is Boltzmann's constant, T is the absolute temperature, and B is the bandwidth.

The blackbody is defined as an idealized material that absorbs all incident energy and reflects none. A blackbody also maintains thermal equilibrium by radiating energy at the same rate as it absorbs energy. A nonideal body will not radiate as much power and will reflect some incident power. The power radiated by a nonideal body can be written as

$$P' = \varepsilon P = \varepsilon kTB = kT_B B \qquad (11.6)$$

where ε is the emissivity, which is a measure of radiation of a nonideal body relative to the ideal blackbody's radiation. Note that $0 \le \varepsilon \le 1$ with $\varepsilon = 1$ for an ideal blackbody. A brightness temperature is defined as

$$T_B = \varepsilon T \qquad (11.7)$$

where T is the physical temperature of the body. Since $0 \le \varepsilon \le 1$, a body is always cooler than its actual temperature in radiometry.

All matter above absolute-zero temperature is a source of electromagnetic energy radiation. It absorbs and radiates energy. The energy radiated per unit wavelength per unit volume of the radiator is given by [13]

$$M = \frac{\varepsilon c_1}{\lambda^3 (e^{c_2/\lambda T} - 1)} \qquad (11.8)$$

where M = spectral radiant existance, $W/(m^2 \cdot \mu m)$

ε = emissivity, dimensionless

c_1 = first radiation constant, 3.7413×10^8 W $\cdot$ $(\mu m)^2/m^2$

λ = radiation wavelength, μm

c_2 = second radiation constant, $1.4388 \times 10^4 \mu m$ K

T = absolute temperature, K

By taking the derivative of Eq. (11.8) with respect to wavelength and setting it equal to zero, one can find the wavelength for the maximum radiation as

$$\lambda_{max} = \frac{2898}{T} \tag{11.9}$$

where T is in kelvin and λ_{max} in micrometers.

Figure 11.10 shows the radiation as a function of λ for different temperatures. It can be seen that the maximum radiation is in the infrared region (3–15 μm). But the infrared remote sensing has the disadvantage of blockage by clouds or smoke.

A typical block diagram of a radiometer is shown in Fig. 11.11. Assuming that the observed object has a brightness temperature T_B, the antenna will pick up a noise power of kT_BB. The receiver also contributes a noise power of kT_RB. The output power from the detector is [14]

$$V_0I_0 = G(T_B + T_R)kB \tag{11.10}$$

where G is the overall gain of the radiometer. V_0 and I_0 are the output voltage and current of the detector. Two calibrations are generally required to determine the system constants GkB and GT_RkB. After the calibration, T_B can be measured and determined from Eq. (11.10).

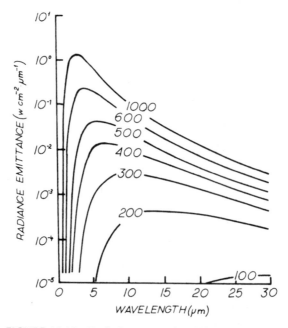

FIGURE 11.10   Radiation curves for different temperatures.

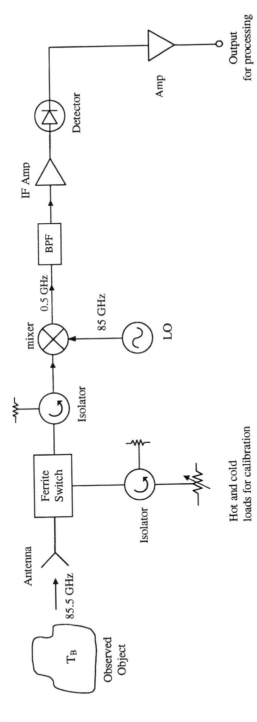

FIGURE 11.11 Radiometer block diagram example.

The major error in this system is due to the gain variation. Such variations occur over a period of 1 s or larger. An error can occur if the measurement is made after the calibrated gain has been changed. The error can be represented by [15]

$$\Delta T_G = (T_B + T_R)\frac{\Delta G}{G} \tag{11.11}$$

As an example, if $T_B = 200$ K, $T_R = 400$ K, and $\Delta G = 0.01G$, then $\Delta T_G = 6$ K. To overcome this error, the Dicke radiometer is used. The Dicke radiometer eliminates this gain variation error by repeatedly calibrating the system at a rapid rate. Figure 11.12 shows a block diagram for the Dicke radiometer. The Dicke switch and low-frequency switch are synchronized. By switching to two positions A and B rapidly, T_{REF} will be varied until V_0 is equal to zero when the outputs from the two positions are equal. At this time, $T_B = T_{REF}$ and T_B is determined from the control voltage V_c. This method eliminates the errors due to the gain variation and receiver noise.

From the brightness temperature measurements, a map can be constructed because different objects have different brightness temperatures. The target classifications need to be validated by truth ground data. Figure 11.13 shows this procedure.

There are many remote sensing satellites operating at different frequencies, from microwave to millimeter wave and submillimeter wave. For example, the SSMI (Special Sensor Microwave Imager) operates at four different frequencies, 19.35, 22.235, 37, and 85.5 GHz [16]. It was designed to monitor vegetation, deserts, snow, precipitation, surface moisture, and so on. Temperature differences of less than 1 K can be distinguished.

The remote sensing satellites have been used for the following applications:

1. *Monitoring earth environments*: for example, ocean, land surface, water, clouds, wind, weather, forest, vegetation, soil moisture, desert, flood, precipitation, snow, iceberg, pollution, and ozone.

2. *Exploring resources*: for example water, agriculture, fisheries, forestry, and mining.

3. *Transportation applications*: for example, mapping road networks, land and aviation scenarios, analyzing urban growth, and improving aviation and marine services.

4. *Military applications*: for example, surveillance, mapping, weather, and target detection and recognition.

11.6 SURVEILLANCE AND ELECTRONIC WARFARE SYSTEMS

Electronic warfare (EW) is the process of disrupting the electronic performance of a weapon (radar, communication, or weapon guidance). It is a battle for the control of electromagnetic spectrum by deliberate means such as interference, jamming with

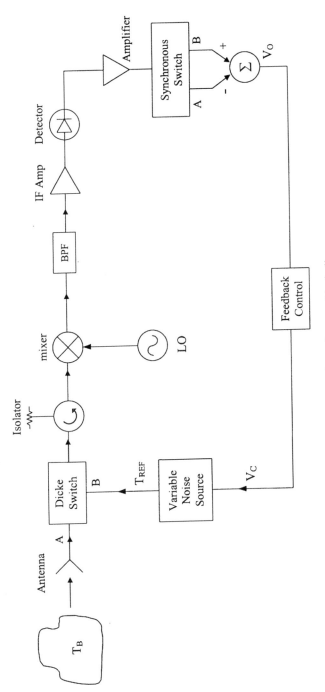

FIGURE 11.12 Dicke radiometer block diagram.

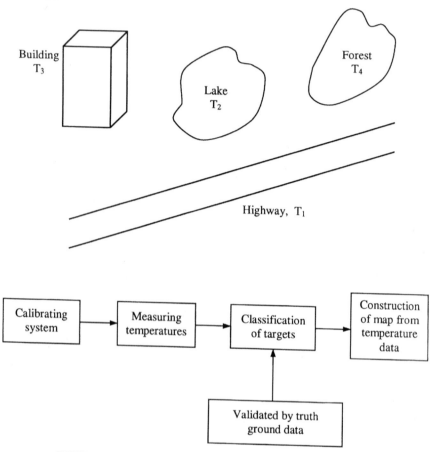

FIGURE 11.13   Procedure for generating remote sensing pictures.

noise, substituting false information (deceptive jamming), and other countermeasures.

Electronic warfare technology can be divided into three major activities: electronic support measure (ESM), electronic countermeasures (ECMs), and electronic counter-countermeasures (ECCMs). Figure 11.14 summarizes these activities [17]. Electronic support measures use a wide-band low-noise receiver to intercept the enemy's communication and radar signals. The intercepted signals will be analyzed to identify the frequency, waveform, and direction. From the intelligence data, the emitter will be identified and recognized. This receiver is also called a surveillance receiver. During peace time, the surveillance is used to monitor military activities around the world. On the battlefield, the findings from ESM will lead to ECM activities. Electronic countermeasures use both passive and active techniques to deceive or confuse the enemy's radar or communication systems. The active ECM system radiates broadband noise (barrage jammers) or deceptive signals (smart

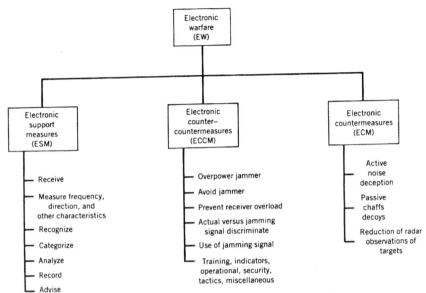

FIGURE 11.14   Modes of electronic warfare [17].

jammers) that confuse or disable the enemy's detection or communication. Passive ECM methods include the use of chaffs or decoys that appear to be targets. A chaff could be a highly reflective material scattered over a large volume to appear as a huge target or multiple targets. The purpose of ECM is to make the enemy's radar and communication systems ineffective. The ECCM actions are taken to ensure the use of the electromagnetic spectrum when the enemy is conducting ECMs. Techniques include designing a receiver with overload protection, using a transmitter with frequency agility, and overpowering a jammer using high-power tubes. The implementation of these techniques requires sophisticated microwave equipment.

11.6.1   ESM System

In the ESM system, the power received from a wide-band receiver is given by the Friis transmission equation, which is similar to the communication system. An example is shown in Fig. 11.15. The received power is

$$P_r = P_t G_t \left(\frac{\lambda_0}{4\pi R}\right)^2 G_r$$

$$= \text{(EIRP) (space or path loss)(receiver antenna gain)} \qquad (11.12)$$

where P_t is the transmitter power from a radar or communication system, G_t is the transmitter gain, λ_0 is the free-space wavelength, R is the range, and G_r is the receiver antenna gain. In many cases, G_t needs to be replaced by G_t^{SL}, which is the

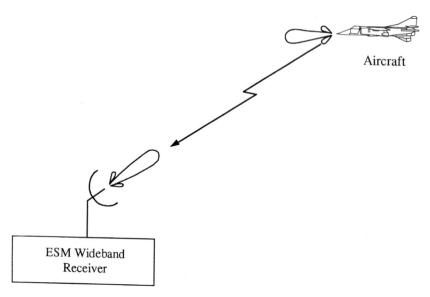

FIGURE 11.15   ESM system.

gain in the sidelobe region since the transmitting signal is not directed to the receiver.

There are many different types of wide-band receivers used for ESM surveillance [17–19]: crystal video receiver, compressive receiver, instantaneous frequency measurement receiver, acousto-optic receiver, and channelized receiver. The crystal video receiver (CVR) consists of a broadband bandpass filter, an RF preamplifier, and a high-sensitivity crystal detector, followed by a logarithmic video amplifier. The approach is low cost and is less complex compared to other methods. The compressive receiver uses a compressive filter when time delay is proportional to the input frequency. By combining with a swept frequency LO signal whose sweep rate $\Delta f / \Delta t$ is the negative of the compressive filter characteristic, a spike output is obtained for a constant input frequency signal. The instantaneous frequency measurement (IFM) receiver uses a frequency discriminator to measure the frequency of the incoming signal. The acousto-optic receiver uses the interaction of monochromatic light with a microwave frequency acoustic beam. The incident light to the acousto-optic medium is diffracted at the Bragg angle, which depends on the wavelength of the incoming signal. The channelized receiver is a superheterodyne version of the crystal video and IFM receivers. It has improved performance but at higher cost and complexity. An example of a channelized receiver is shown in Fig. 11.16 covering 2–18 GHz incoming frequency range [17]. The incoming signal is downconverted to lower IF frequencies. By using a group of decreasing bandwidth filter banks (2 GHz, 200 MHz, and 20 MHz) and the associated downconverters, the signal will appear from one of the detectors. From the detector and the switching information, one can determine the frequency of the incoming signal.

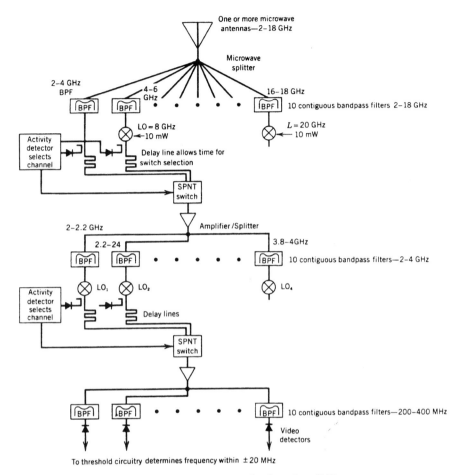

FIGURE 11.16 Channelized receiver [17].

11.6.2 ECM Systems

In ECM systems, the radar equation and Friis transmission equation can be used to describe the jamming scenarios. Two common scenarios are the self-screening jammer (SSJ) and a CW barrage stand-off jammer (SOJ) screening the attack aircraft as shown in Fig. 11.17 [17]. Considering the SSJ case first (Fig. 11.17a), the interrogating radar is jammed by a jammer located at the target. The jammer wants to radiate power toward the radar to overwhelm the target return. Effective jamming would require the jammer-to-signal (J/S) ratio received by the radar to exceed 0 dB. From the radar equation (7.12), we have the returned signal from the target given by

$$S = P_r = \frac{P_t G_r^2 \sigma \lambda_0^2}{(4\pi)^3 R^4} \tag{11.13}$$

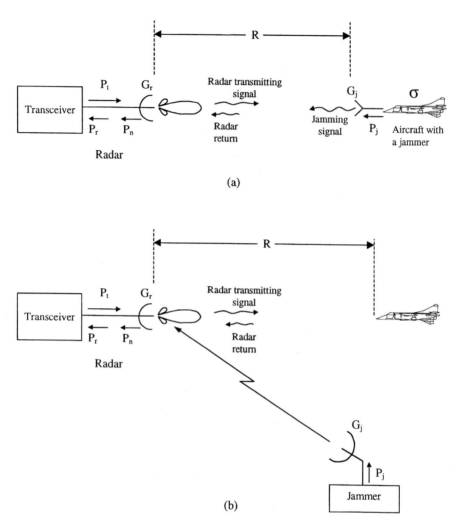

FIGURE 11.17 Two different jamming scenarios: (a) self-screening jamming (SSJ) and (b) stand-off jamming (SOJ).

The jammer transmits a power level P_j from its antenna with a gain G_j. The jammer has the advantage of one-way loss, and the Friis transmission equation (8.21) is used to calculate the received signal at the radar:

$$J = P_n = P_j G_j G_r \left(\frac{\lambda_0}{4\pi R}\right)^2 \tag{11.14}$$

The J/S ratio can be found by Eqs. (11.13) and (11.14) as

$$\frac{J}{S} = \frac{P_n}{P_r} = \frac{P_j}{P_t} \frac{4\pi R^2 G_j}{G_r \sigma} \tag{11.15}$$

The above equation is valid only when the jammer and radar have the same bandwidth. In practice, the jammer needs to radiate over a broader bandwidth than the radar to be sure that the returned signal is blanked. If B_r is the radar bandwidth and B_j is the jammer bandwidth, the power of the jammer will spread out the bandwidth of B_j. Equation (11.15) is modified to account for the bandwidth effects. We have

$$\frac{J}{S} = \frac{P_n}{P_r} = \frac{P_j}{P_t} \frac{4\pi R^2 G_j}{G_r \sigma} \left(\frac{B_r}{B_j}\right) \tag{11.16}$$

For the stand-off scenario (Fig. 11.17b), the jammer is situated on another location other than the target. The radar is aimed at the target, and the jammer is located off the radar's main beam. The jamming signal received by the radar is

$$J = P_n = P_j G_j \cdot G_r^{SL} \left(\frac{\lambda_0}{4\pi R_j}\right)^2 \tag{11.17}$$

where R_j is the distance between the jammer and the radar and G_r^{SL} is the gain in the sidelobe region of the radar antenna. From Eqs. (11.13) and (11.17), the J/S ratio is

$$\frac{J}{S} = \frac{P_n}{P_r} = \frac{P_j G_j G_r^{SL} (4\pi) R^4}{P_t G_r^2 \sigma R_j^2} \tag{11.18}$$

If the bandwidths of the radar and jammer are different, Eq. (11.18) becomes

$$\frac{J}{S} = \frac{P_j G_j G_r^{SL} (4\pi) R^4}{P_t G_r^2 \sigma R_j^2} \left(\frac{B_r}{B_j}\right) \tag{11.19}$$

Example 11.1 A 3-GHz tracking radar has an antenna gain of 40 dB, a transmitter power of 200 kW, and an IF bandwidth of 10 MHz. The target is an aircraft with a radar cross section of 5 m^2 at a distance of 10 km. The aircraft carries an ECM jammer with an output power of 100 W over a 20-MHz bandwidth. The jammer has an antenna gain of 10 dB. Calculate the J/S ratio.

Solution

$$B_r = 10 \text{ MHz} = 10^7 \text{ Hz} \qquad B_j = 20 \text{ MHz} = 2 \times 10^7 \text{ Hz}$$
$$G_r = 40 \text{ dB} = 10^4 \qquad\qquad \sigma = 5 \text{ m}^2$$
$$G_j = 10 \text{ dB} = 10 \qquad\qquad R = 10 \times 10^3 \text{ m}$$
$$P_t = 200 \text{ kW} = 2 \times 10^5 \text{ W} \qquad P_j = 100 \text{ W} = 10^2 \text{ W}$$

From Eq. (11.16), the J/S ratio is given as

$$\frac{J}{S} = \frac{P_j \, 4\pi R^2 G_j}{P_t \quad G_r \sigma} \left(\frac{B_r}{B_j}\right)$$

$$= \frac{10^2 \times 4\pi \times (10 \times 10^3)^2 \times 10}{2 \times 10^5 \times 10^4 \times 5} \left(\frac{10^7}{2 \times 10^7}\right)$$

$$= 62.83 \text{ or } 17.98 \text{ dB}$$

PROBLEMS

11.1 In an active radio navigation system, an aircraft transmits a pulse-modulated signal at f_1. The return pulses at f_2 are delayed by 0.1 msec with a measurement uncertainty of 0.1 μsec. Determine (a) the distance in kilometers between the aircraft and the station and (b) the uncertainty in distance in meters.

11.2 A passive radio navigation system transmits a signal to an aircraft. The signal arrives 1 msec later. What is the distance between the aircraft and the station?

11.3 A ship receives signals from three radio navigation stations, as shown in Fig. P11.3. The coordinates for the ship and the three stations are (x, y), (x_1, y_1), (x_2, y_2), and (x_3, y_3), respectively. The receiver clock error is ε. Derive three equations used to determine (x, y) in terms of the signal travel times and ε.

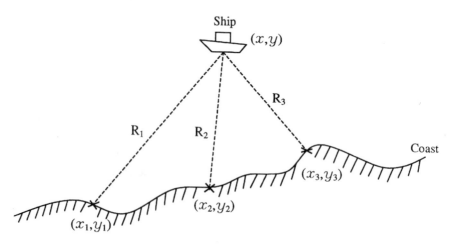

FIGURE P11.3

11.4 The DBS receiver shown in Fig. P11.4 consists of an RF amplifier, a mixer, and an IF amplifier. The RF amplifier has a noise figure of 4 dB and a gain of

15 dB. The mixer has a conversion loss of 6 dB, and the IF amplifier has a gain of 35 dB and a noise figure of 8 dB. Calculate the overall noise figure and gain.

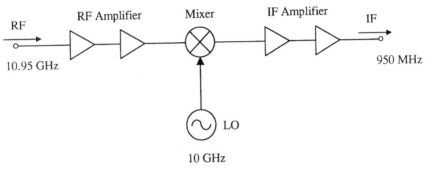

FIGURE P11.4

11.5   The RFID system shown in Fig. P11.5 consists of a reader and a passive tag. The reader transmits a signal at 915 MHz with a power level of 100 mW. The reader antenna has a gain of 10 dB. The tag is located 1 ft away from the reader with an antenna gain of 3 dB. The signal received by the tag is converted to 1830 MHz by a diode frequency doubler. The conversion efficiency is 10%. Assuming that the antenna is in the far-field region, determine the power level P_0 after the conversion. This power is transmitted back to the reader through a tag antenna.

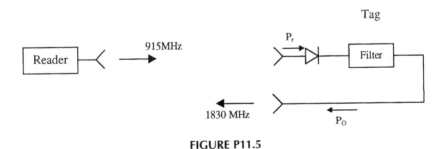

FIGURE P11.5

11.6   A 10.5-GHz police radar detector is used for surveillance. The detector consists of a $p-i-n$ diode as an RF modulator followed by a detector, as shown in Fig. P11.6a. The police radar transmits a 10.5-GHz CW signal. Draw the waveform after the $p-i-n$ diode (point A) and the waveform for the output at the detector (point B). The bias to the $p-i-n$ diode is shown in Fig. P11.6b, and the pulse repetition rate is in the kilohertz region. Explain how the system works.

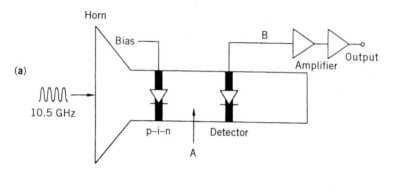

(a)

10.5 GHz

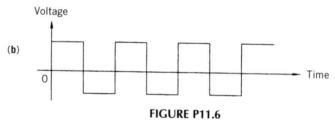

(b)

FIGURE P11.6

11.7 A 10-GHz tracking radar has an antenna gain of 30 dB, a transmitting power of 100 kW, and an IF bandwidth of 20 MHz. The target has a cross section of 10 m^2 and carries an ECM jammer with an output power of 100 W over a 30-MHz bandwidth. The jammer has an antenna gain of 10 dB. The target is located at a distance of 5 km from the radar. Calculate the J/S ratio.

11.8 In Problem 11.7, if the same jammer is a stand-off jammer located at a distance of 5 km from the radar, what is the J/S ratio? The radar sidelobe level at the stand-off jammer direction is 20 dB down from the peak level.

REFERENCES

1. I. A. Getting, "The Global Positioning System," *IEEE Spectrum*, Dec. 1993, pp. 36–47.
2. T. Logsdon, *The Navstar Global Positioning System*, Van Nostrand Reinhold, New York, 1992.
3. L. H. Eriksson and S. Broden, "High Performance Automotive Radar," *Microwave J.*, Vol. 39, No. 10, pp. 24–38, 1996.
4. H. Bierman, "Personal Communications, and Motor Vehicle and Highway Automation Spark New Microwave Applications," *Microwave J.*, Vol. 34, No. 8, pp. 26–40, 1991.
5. H. H. Meinel, "Commercial Applications of Millimeter-Wave History, Present Status, and Future Trends," *IEEE Trans. Microwave Theory Tech.*, Vol. 43, No. 7, pp. 1639–1653, July 1995.

6. R. Dixit, "Radar Requirements and Architecture Trades for Automotive Applications," *IEEE Int. Microwave Symp. Dig.*, June 1997, pp. 1253–1256.

7. K. W. Chang, H. Wang, G. Shreve, J. G. Harrison, M. Core, A. Paxton, M. Yu, C. H. Chan, and G. S. Dow, "Forward-Looking Automotive Radar Using a W-band Single-Chip Transceiver," *IEEE Trans. Microwave Theory Tech.*, Vol. 43, No. 7, pp. 1659–1668, 1995.

8. S. Yoshida, K. Satoh, T. Miya, T. Umemoto, H. Hirayama, K. Miyagaki, and J. Leong, "GaAs Converter IC's for C-Band DBS Receivers," *IEEE J. Solid-State Circuits*, Vol. 30, No. 10, pp. 1081–1087, 1995.

9. P. M. Bacon and E. Filtzer, "What is Behind DBS Services: MMIC Technology and MPEG Digital Video Compression," *IEEE Trans. Microwave Theory Tech.*, Vol. 43, No. 7, pp. 1680–1685, 1995.

10. Data Sheet for AKD 12000, ANADIGICS Inc., Warren, NJ.

11. J. Eagleson, "Matching RFID Technology to Wireless Applications," *Wireless Systems* May 1996, pp. 42–48.

12. D. D. Mawhinney, "Microwave Tag Identification Systems," *RCA Review*, Vol. 44, pp. 589–610, 1983.

13. K. Szekielda, *Satellite Monitoring of the Earth*, John Wiley & Sons, New York, 1988.

14. D. M. Pozar, *Microwave Engineering*, 2nd. ed., John Wiley & Sons, New York, 1998.

15. F. T. Ulaby, R. K. Moore, and A. K. Fung, *Microwave Remote Sensing: Active and Passive*, Vol. I, Adison-Wesley, Reading, MA, 1981.

16. C. M. U. Neale, M. J. McFarland, and K. Chang, "Land-Surface-Type Classification Using Microwave Brightness Temperatures from the Special Sensor Microwave/Imager," *IEEE Trans. Geosci. Remote Sens.*, Vol. 28, No. 5, pp. 829–838, 1990.

17. E. A. Wolff and R. Kaul, *Microwave Engineering and Systems Applications*, John Wiley & Sons, New York, 1988.

18. J. Tsui, *Microwave Receivers with Electronic Warfare Applications*, John Wiley & Sons, New York, 1986.

19. J. Tsui and C. H. Krueger, "Components for Surveillance and Electronic Warfare Receivers," *Handbook of Microwave and Optical Components*, Vol. 1, K. Chang, Ed., John Wiley & Sons, New York, 1989.

Index